Handbook of Watch and Clock Repairs

HANDBOOK OF

Watch & Clock Repairs

REVISED EDITION

H.G. Harris

BARNES & NOBLE BOOKS
A DIVISION OF HARPER & ROW, PUBLISHERS
New York, Cambridge, Philadelphia, San Francisco
London, Mexico City, São Paulo, Sydney

HANDBOOK OF WATCH AND CLOCK REPAIRS. Copyright © 1961, 1963, 1972 by H. G. Harris. All rights reserved. Printed in the United States of America. No part of this book may be used or reproduced in any manner whatsoever without written permission except in the case of brief quotations embodied in critical articles and reviews. For information address Harper & Row, Publishers, Inc., 10 East 53rd Street, New York, N.Y. 10022. Published simultaneously in Canada by Fitzhenry & Whiteside Limited, Toronto.

First BARNES & NOBLE BOOKS edition published 1984.

Library of Congress Cataloging in Publication Data

Harris, H. G. (Henry Gordon)
    Handbook of watch and clock repairs

    (Everyday handbook)
    Includes index.
    1. Clocks and watches—Repairing and adjusting.
I. Title.
[TS547.H34   1984]   681.1'13   83-48352
ISBN 0-06-463591-0 (pbk.)

84 85 86 87 88 10 9 8 7 6 5 4 3 2 1

Contents

CONTENTS

Preface

THERE have been many books written on the repair of watches and clocks but the majority have been intended for the serious horologist and the apprentice.

Little has been done to publicize horology as a hobby.

Many will say that to do such a thing is inadvisable without proper training. But what of the present-day hobbyists who are self-taught and whose interest in their subject is such that their knowledge and skill often surpass those whose full-time occupation it is?

I have known men who, since their youth, have been interested in radio. Their knowledge and skill today are extensive and although not qualified in the official sense, nevertheless they have become authorities on their subject. Similar remarks can be directed towards other occupations, and so it is with watch and clock repairing.

In writing this book, I have assumed that the reader has no knowledge of the subject and I have endeavoured, therefore, to concentrate on the basic principles rather than advanced work. An attempt has also been made to show the beginner that quite a lot of practical work can be done with limited equipment and a small initial outlay.

In the course of overhauling a movement the beginner will frequently be confronted with a job beyond his ability and which requires the use of equipment not in his possession. The appendix at the back of the book will guide the reader on how to go about sending work to an outside repairer. Once having established a contact, no job should be too big or too difficult to tackle.

A little practice at home dismantling and assembling some old movements will quickly introduce confidence and provide the reader with the light touch necessary when working on watches.

Appreciation and thanks are extended to Messrs Parechoc S.A., Le Sentier, Switzerland (manufacturers of the Kif Flector), The Universal Escapement Ltd., La Chaux-de-Fonds, Switzerland (manufacturers of the Incabloc), and to Erismann Schinz Ltd., Le Neuveville, Switzerland (manufacturers of the Monorex) for supplying me with detailed information and drawings of their shock absorbers.

A special word of thanks is given to Messrs Baume & Co. Ltd, 50 Hatton Garden, London (Longine watches) and Smiths Clocks & Watches Ltd, Sectric House, Cricklewood, London, both of whom have been most helpful in supplying information and drawings.

It is worthy of note here that in my approach to the industry, I found an unexpected enthusiasm to help when it was known that the book was to be a hobby book rather than a textbook. It was considered that such a book was badly needed and all concerned wished it every success.

Last, but by no means least, my special thanks go to Mr S. Pleasants who has done so much in producing original drawings and preparing illustrations for publication, frequently, I might add, having to alter them as a result of a change in the original manuscript. For him it was a nightmare gallop keeping pace with the typewriter keys.

CHAPTER ONE

Workbench and tools

WATCHES and clocks can be overhauled and simple repairs carried out without the need of an elaborately equipped workshop. To begin with, a small rigid table with adequate lighting will provide the workbench, and the tools can be limited initially to a selected few to cover general work. More specialized tools can be obtained later if and when the need arises.

However, it is the intention here to describe typical workshop conditions and then leave you free to modify these arrangements to suit your requirements.

A workbench is required, or alternatively a shelf firmly secured to a wall.

The working surface of the bench should be at least 3 ft. from the ground (Fig. 1). This will enable close work to be carried out without bending low, resulting in greater comfort and increased control. The length of the bench top needs to be about 3 ft. 6 in. to permit working with both elbows spread out. The width should be between 18 and 20 in. to allow space behind the work in hand for small tools and accessories in use.

Strips of wood about $2\frac{1}{2}$ in. wide placed on edge on the top of the bench along the two ends and along the back prevent items being knocked to the floor.

A few drawers built under the top for storage of work, tools, materials, etc., completes the bench.

It is essential that all work be conducted away from dirt or dust because the smallest piece of foreign matter in a watch can cause the movement to stop, and dust will soak up the oil and cause the movement to dry out.

Sometimes it is quicker and cheaper to secure a shelf to a wall. A strong, well made shelf securely fitted is much preferred by many repairers.

Preferably the bench should be situated against a window facing north. The reflected light obtained is softer and more suited to this type of work.

If an adjustable electric lamp is mounted on the bench top or secured to the wall in the case of a shelf, the lamp can be pulled down close to the work and there should be no fear of eye strain.

Having provided the workbench and the lighting, attention

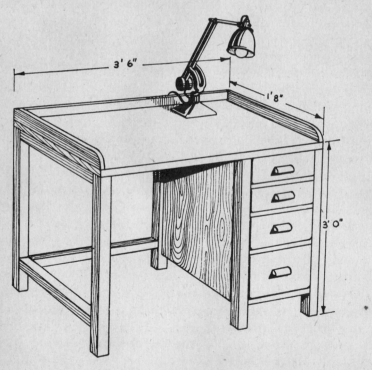

Fig. 1. Workbench.

must now be given to tools. The following list is a guide to the tools that need to be obtained to start with and to those that can be purchased when a little more experience has been gained and the range of work being undertaken has been widened.

INITIAL TOOL KIT

1 pair flat-nose pliers 4 in.
1 screwdriver 7 in.
1 set watch screwdrivers
1 eye-glass 2-in. focus
1 pair fine tweezers
1 pair heavy tweezers
2 oilers
1 oil cup (watch oil)
1 oil cup (clock oil)
2 pairs hand removing levers
1 pith holder

2 cleaning brushes (medium and soft)
1 pin-vice
1 oilstone, fine/medium
handles
broaches
rat-tail files
pillar files – fine, medium and coarse – 4 and 6 in.
1 vice, $2\frac{1}{2}$ to 3-in. jaws
1 watchmaker's hammer

SUPPLEMENTARY TOOL KIT

1 pair round-nose pliers 4 in.
1 pair snipe-nose pliers 4 in.
1 pair brass-faced flat-nose pliers 4 in.
1 pair top-cutting nippers 4 in.
1 screwdriver 12 in.
1 double lens eye-glass $\frac{1}{4}$-in. focus
1 eye-glass 3-in. focus

1 round-face hammer
1 brass-faced hammer
1 watchmaker's spirit lamp
1 blueing pan
1 graduated steel stake
1 flat burnisher
glass dust covers
1 pair turns with accessories
2 clock spring clamps

Too much emphasis cannot be placed on the need to buy good quality tools. It is better to start with a few good ones and slowly build up a kit, than to buy poor quality tools that must inevitably lead to repair work of a low standard. Make your purchases from firms who specialize in this equipment and who thoroughly understand your requirements.

The application of special tools will be dealt with in the appropriate chapters, but a few hints on the use and maintenance of the more common ones may prove helpful at this stage.

Files. The two most useful types of file for watch and clock repairing are pillar files and rat-tail files (Fig. 2). Useful sizes are 4 in. and 6 in. in coarse, medium and fine cuts.

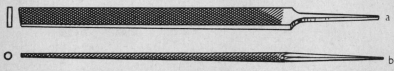

Fig. 2. Files. (a) Pillar (b) Rat-tail.

Files are cutting tools and must be treated as such. The teeth are shaped like those of a saw and consequently cut in one direction only.

Soft metals such as brass need new files but steels can be cut better with a part worn file where the tips of the teeth have been worn down. It follows then that new files commence their life by being used for brass and are then passed over for use on steels.

It is sometimes more economical to use a double-sided file such as a pillar file, for both metals, in which case the side used for brass is marked by passing a piece of white chalk across the teeth.

It is essential that you are able to file accurately and therefore a little practice may prove worth while.

Place a piece of brass rod horizontally in the vice. Hold the pillar file by the handle in the right hand. Place the file on the brass and place the forefinger and thumb of the left hand on the end of the file.

A forward stroke is now made with just sufficient downward pressure for the file teeth to cut the metal.

A flat surface can best be obtained by keeping the file horizontal. To do this the stroke is made with a light downward pressure on the file handle and a heavier downward pressure on the file tip.

As the centre of the file nears the brass, so the two pressures are adjusted until, when the file is equidistant over the brass the two pressures are the same.

As the stroke continues, the two downward pressures are progressively reversed until, when the file has reached the end of its stroke, the heavier downward pressure is being made on the handle and the lighter downward pressure on the tip.

A little practice and it will be found that the knack of maintaining these varying pressures will not be difficult.

Pillar files usually have both edges smooth thus enabling one face of a step to be filed without removing metal from the adjacent face.

Quite frequently taper pins will have to be made by filing; this is known as 'pin-filing'.

A small block of fibre or hard wood is needed having in one face a number of grooves of varying depths. The block is placed in the vice with the grooves uppermost. A piece of selected brass rod or wire is placed in the jaws of the pin-vice (Fig. 3), the protruding length dependent on the length of the pin required.

The wire is laid in one of the grooves and held at a slight downward angle. The groove selected is that which allows the wire to just stand proud of the surface of the block.

A fine pillar file is selected and a light forward stroke is made. The right hand can control the file better if the forefinger is

Fig. 3. Pin-vice.

straight and its tip resting on the side of the file. At the same time the pin-vice is rotated between the forefinger and thumb of the left hand causing the wire to revolve against the direction of the file.

When the stroke is completed, the file is drawn back, lightly resting on the wire to keep the wire in the groove, and the pin-vice is rotated in the opposite direction in readiness for the next stroke.

A little practice and synchronization of movement will be achieved.

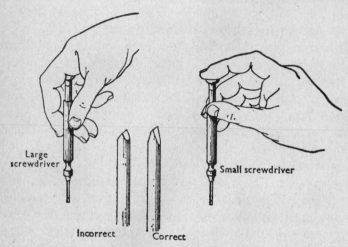

Fig. 4. Correct methods of holding watchmaker's screwdrivers; and (*centre*) screwdriver blade.

Watch screwdrivers. These are best purchased as a set of three or four in a box. The correct method of holding them is shown in figure 4.

With constant use screwdriver blades need re-shaping. This is done by using a smooth file. The blade of the screwdriver should be well blunted and the taper kept long (Fig. 4). If the taper is short and the blade end is sharp, the screwdriver will tend to rise out of the screwdriver slot and damage the screw head.

Oilers. These can be purchased quite cheaply in attractive plastic cases, but on the other hand they are simple to make which sometimes proves more satisfying.

Two sizes are required. A small one for jewel holes and a larger one for general work.

The small one can be made from a sewing needle. Heat the point to a blue, file two flats opposite each other, place the needle on the vice and tap out the end using the round-face hammer. Finish off with a small oilstone. Figure 5 shows an enlarged view of the shaped end.

The larger one can be made from a short length of steel wire.

Long slender handles are now needed. These can be made from lengths of wooden rod rounded off at each end.

The method of transferring oil from the oil bottle to the movement is accomplished in two stages. First, a large watch screwdriver is dipped into the bottle of oil and one drop of oil is placed in an oil cup. The tip of the oiler is then placed in the oil resulting in a small quantity of oil being deposited on the oiler.

The tip of the oiler is then allowed to touch the part to be oiled and if the oiler is correctly shaped the oil will be transferred by capillary action.

Eye-glasses. The list of tools includes three sizes of eye-glasses, a $\frac{1}{4}$-in. focus double lens, a 2-in. focus single lens (Fig. 6) and a 3-in. focus single lens. The $\frac{1}{4}$-in. is for examining very fine work such as jewel holes in watches. The 2-in. is for general use when working on watches, and the 3-in. is for clock work.

The lens holder is fitted into the socket of the eye where it should remain without discomfort leaving both hands free to work.

The lens holders vary slightly in diameter and thickness and it

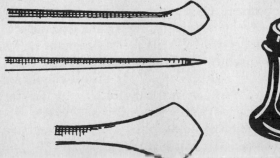

Fig. 5. Oilers.

Fig. 6. Eye-glass.

is sometimes more satisfactory to try a few glasses for fit before making a purchase.

If after continual use it is found that the inner face of the lens tends to steam up, two or three holes about $\frac{1}{8}$ in. diameter drilled through the side of the holder will cure the trouble.

Some repairers prefer to fit head-wires to their eye-glasses. At least this does prevent the eye-glass from falling at a moment when a delicate operation is in progress.

A piece of spring steel wire is looped at one end to hold the eye-glass and the remainder is curved to fit the shape of the head. The end should reach the back of the head and then be bent into a small loop to prevent the end from digging in.

For those who wear spectacles, eye-glass holders can be purchased that will fit the spectacle frames. Your optician will be the best person to give advice in this respect.

Tweezers. Apart from eye-glasses, tweezers are possibly the most used tools, particularly in watch repairing. Tool manufacturers produce wide ranges of tweezers of many shapes, sizes, qualities and uses and it is therefore sometimes difficult to know how to make a selection.

It is advisable to start with two pairs of good quality tweezers for general use. One for light work and the other a little more robust.

Earlier in this chapter advice was given against buying cheap tools. This applies particularly to tweezers.

The slightest tendency to twist or reluctance to grip properly may result in the part snapping out and being lost or damaged.

A good pair of tweezers will pick up a hair between the points from a piece of glass. It should be possible to place a piece of thin metal between the points and apply plenty of pressure without the points curling outward. If they do curl outward (Fig. 7), then the points need trimming. This is best done by using a small oilstone. If the outward curve is considerable, then the points must be carefully bent inward using the small flat-nose pliers.

Pliers (Fig. 8). If the serrations on the inner face of the pliers are too coarse, the tops of the serrations can be removed by filing.

The brass-faced pliers are for handling delicate work without damage or marking.

Top cutting nippers (Fig. 9). These are used for cutting lengths of wire. Keep the cutting edges sharp by using a medium cut pillar file.

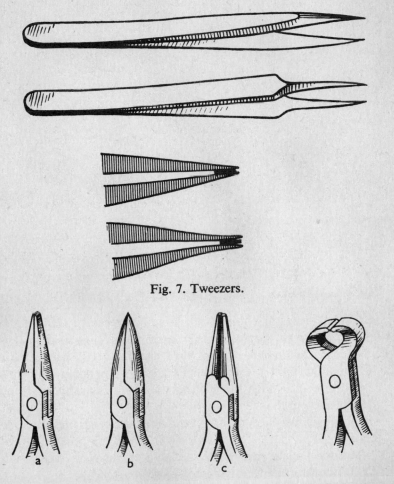

Fig. 7. Tweezers.

Fig. 8. Pliers. (a) Flat-nose (b) Snipe-nose (c) Round-nose.

Fig. 9. Top-cutting pliers.

Hand-removing levers. These are essential for removing hands and are easily made. Two pairs are needed, one fine pair for watches and a larger pair for clocks.

Make the levers from brass as shown in figure 10.

Pith holder. A round metal container about 1½ in. diameter

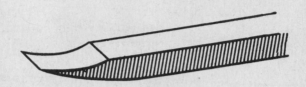

Fig. 10. Hand lifting lever.

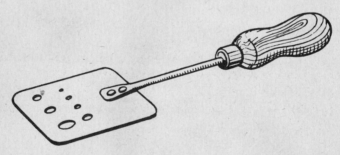

Fig. 11. Blueing pan.

and 1½ in. deep is required. Place some pieces of lead in the bottom and melt them down. A layer about ¼ in. is all that is needed, enough to make it firm when standing on the bench. Pack the container tight with pith sticks about 1¼ in. long. The ends of tweezers and watch screwdrivers are kept clean by prodding them into the pith.

Tools such as files, gravers, etc., must have handles fitted to them.

Heat the handle end of the tool to a cherry red making sure that the cutting end is not affected by the heat.

Burn the tool into the handle almost to the desired position. Hold the tool vertically with the handle downward and strike the vice a few hard blows with the tool handle. This should result in the handle being firmly in position.

Broaches. When broaching a hole do not use force. Small broaches can be held in a pin-vice and rotated between the forefinger and thumb. Larger broaches need thin wooden handles. When cutting with a broach plenty of lubrication and frequent clearing of broach cuttings is necessary.

Blueing pan. This is used for blueing polished steel parts The pan is a piece of brass sheet measuring approximately $1\frac{1}{2}$ in. $\times 1$ in. and drilled with a row of graduated holes at one end.

To the other end is riveted a short length of steel rod over which a wooden handle is driven (Fig. 11).

The parts to be blued are placed on the pan which is then passed through a spirit flame.

As the temperature of the steel parts increases, the surface colour of the steel changes to blue.

The pan is removed from the flame and the parts are either dropped into oil or they are allowed to cool off and given a coat of colourless lacquer.

Graduated Stake. (Fig. 12). This is a flat steel block, hardened

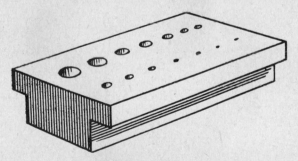

Fig. 12. Graduated stake.

and polished, and drilled with holes of different sizes to suit the work in hand. The under-side is stepped so that it can be held between the jaws of a vice.

Watchmaker's Hammer (Fig. 13). These hammers can be purchased with heads of steel or brass. The steel head is a general purpose hammer, but when the work is more delicate and the risk of damage greater, a brass head is used.

Finally, a useful set of tools (Fig. 14) is supplied by Smiths Clock and Watch Division, London, England.

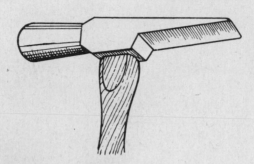

Fig. 13. Watchmaker's hammer.

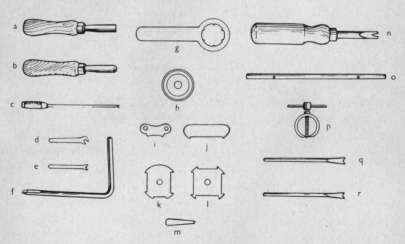

Fig. 14. Smith's set of tools.
(a) Case nut spanner. (b) Alarm stop spanner. (c) 'C' clip remover. (d) Balance screw key. (e) Hairspring collet adjuster. (f) Bezel remover. (g) De Luxe 12 ligne case opener. (h) Suction case opener. (i) De luxe 8¾ ligne case opener. (j) Yachting timer opener. (k) Empire watch case opener. (l) De luxe dustproof case opener. (m) Empire crown key. (n) Special screwdriver. (o) Hexagon nut spanner (8 × 10 BA). (p) Circular hand extractor. (q and r) Pair of hand lifting levers.

CHAPTER TWO

Materials

Oil. The lubrication of watches and clocks is done by oil refined specially for the purpose. Clock oil has a slightly greater viscosity than watch oil. Both are supplied in small bottles of convenient size.

In Chapter 5 we shall be discussing the oiling of a watch movement and when you realize how very little oil a complete watch needs, you will then appreciate that even a small bottle will last a very long time.

When ordering your oil, be it for watches or clocks, buy the very best. Poor quality oil soon thickens and you will find yourself dismantling and cleaning the movement all over again.

Make sure it is watch or clock oil. Don't be misled into thinking that light machine oil will do, no matter how superior the quality may be.

If the viscosity of the oil is too great for the working parts, the resistance is going to effect the time-keeping.

Pegwood. Sticks of pegwood are sold in bundles about 6 in. long. They are used mainly for cleaning out pivot holes. This is done by shaving one end of a stick to a fine point, inserting the point through the jewel hole and lightly twisting between the finger and thumb. Any dirt in the hole will become embedded in the pegwood. Great care must be taken to ensure that the point does not break off in the pivot hole. Until experience is gained it is advisable to examine the pivot hole after this operation to make sure it is clear. If a piece does break off it can be removed by inserting the pegwood from the other side.

Methylated spirit. This is used as fuel in the watchmaker's spirit lamp. When burning, it gives off a clean smokeless flame and does not blacken or tarnish articles heated in it. This is essential when tempering steel because of the necessity of watching the colour change.

Benzine and Gasoline. As a cleaning agent for watch and clock movements, benzine is undoubtedly the best. It can be purchased from a chemist or drug store. A good alternative to benzine is gasoline, particularly that sold as fuel for cigarette lighters. Both fluids have a high rate of evaporation and should therefore be kept in air-tight containers.

WARNING : *Both benzine and gasoline are highly inflammable and must therefore be kept well away from naked flames. Under no circumstances have them standing near when the watchmaker's spirit lamp is in use.*

Cleaning fluids. As an alternative to using benzine or gasoline there are many proprietary cleaning fluids available on the market. Providing they are supplied by a reputable supplier these cleaning fluids have much to commend them.

Polishing powders. These are commonly known as crocus powder, red-stuff and rouge, the only difference being their fineness of grain. Rouge is the coarsest and crocus powder the finest. The powder is mixed with oil to produce a cutting paste. The medium grade is that most commonly used on watches.

Chalk. This is used on the cleaning brushes for cleaning the movements and to maintain the cleanliness of the brushes themselves. A convenient form is billiard chalk. It is applied by stroking the brush over the chalk a few times.

Pith. The cleaning of watch pivots and similar parts is done by using pith. It is the pith from an elder tree that has been dried, peeled and prepared in sticks.

Emery sticks. Two pieces of hard wood approximately 5 in. × 2 in. × $\frac{1}{2}$ in. planed flat both sides are required. Four grades of emery paper ranging from very fine to medium are cut into 2 in. strips. One strip from each grade is glued to the wood. These emery sticks, as they are now called, are very useful for reducing or smoothing steel or brass surfaces.

Dial enamel. This is used for repairing white enamel dials. It has a low melting point and sets hard with a glossy surface.

CHAPTER THREE

The turns and their uses

THERE are two methods of turning. One is to use a lathe and the other is by means of turns.

Turns are a simple device. Motive power is produced by a hand-operated bow looped once around a ferrule, and the graver (cutting tool) is held in the other hand. A little practice is necessary to acquire the knack of synchronizing these two movements and after that, simple turning can be done with a high degree of accuracy.

The lathe has many advantages over the turns. It is not hand operated which means both hands are free for the work, and it is far more comprehensive in the number of operations and functions it can perform. Being a precision machine it is easier to achieve the same degree of accuracy.

The turns consist of two centres and a handrest with a steel or brass beam passing through them (Fig. 15). One of the centres is secured to the beam at one end, and it is here that the turns are held in a vice. The handrest and remaining centre are free to slide and can be locked to the beam in any position.

Turning. The work to be turned is placed on a turning arbor (Fig. 16), or in an adjustable ferrule (Fig. 17). In both cases the assembly is supported between the centres.

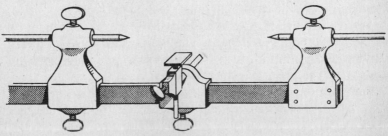

Fig. 15. The turns.

The handrest is brought close to the work and locked in position.

The string of the bow is passed once around the ferrule and the

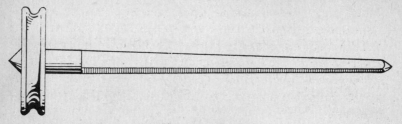

Fig. 16. Arbor.

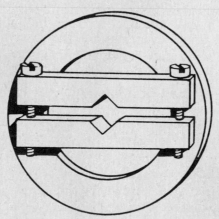

Fig. 17. Adjustable ferrule.

bow held in the left hand. A turning graver is held in the right hand and is placed on the handrest close to the work.

At each downward stroke of the bow the graver is brought up to the work and a cut is made. The bow is then moved upward and the graver is moved away from the work just enough to give a clearance. The process is then repeated.

Turning arbors and adjustable ferrules are supplied in a wide range of sizes. The bows are supplied in a few sizes and there is a choice of materials from which the bow-string is made.

When using a turning arbor, a hole is drilled in the metal from end to end which is then broached out smooth. The tapered hole will now push tightly on to the arbor.

Holding the work by an adjustable ferrule involves the selection of a ferrule of suitable size and then clamping it to the work.

The gravers are made from square or diamond section hardened steel. The cutting ends are ground as shown in figure 18. The graver at (a) is used for roughing down the metal, (b) is for finishing off a square shoulder, cuts being made by the side of the tool as well as the point, and (c) is used for turning radiused shoulders.

After grinding, the face is levelled and smoothed on an oilstone, and finally polished on an Arkansas stone. The side faces are then held flat on the stone and drawn carefully across to remove any burrs.

Each graver should be fitted with a wooden handle as described in Chapter 1.

Having secured the work on an arbor or in a ferrule, it is now placed between the centres and the centres are brought up into

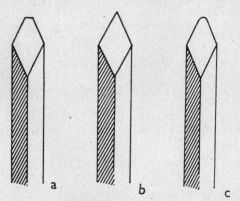

Fig. 18. Gravers.

position and locked. There should be no end float of the work. Apply a little oil at each centre.

Adjust the handrest to bring it close to the work. The exact distance will come with experience. If the distance is too great the graver point will tend to drop and cause chattering. If the hand-rest is too close, there will be insufficient space to rest the graver and operate freely.

The height of the handrest must now be adjusted. Cutting is done by the point of the tool which has to be in line with the

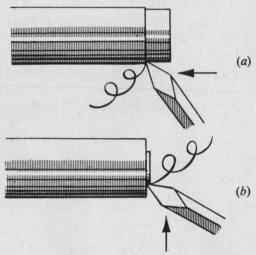

Fig. 19. Turning.

centre of the work. If the graver is too high it will not cut, and if it is too low it will tend to lift the work and be drawn under. It is during the downward stroke of the bow that the cuts are made.

Let us assume we have a piece of ¼ in. diameter brass rod ¼ in. long that has been drilled through its length and broached to fit a suitable size arbor.

The work is mounted between the centres and the turns made ready for use.

First let us reduce the diameter. The graver is positioned as shown in figure 19 (a) and must be held quite firmly. A downward movement of the bow is made and the graver is allowed to just touch the work at the same time moving slightly to the left. If the work is not true an indication will be given by the graver not making a complete cut but only removing metal from the high spots.

The bow is given an upward stroke and the graver is moved away from the work by means of very slight finger pressure.

Carrying on from where the previous stroke finished, the operation is repeated until sufficient length of the material has been cut.

The graver is then returned to the beginning and the metal is further reduced in diameter.

This process is repeated until the work is revolving completely true and the diameter has been reduced to the predetermined size.

Now to turn the face. The graver is held as shown in figure 19 (*b*) and the process of cutting repeated, this time moving the graver fore and aft.

Brass requires a higher cutting speed than steel and therefore a longer bow is needed.

Bows are supplied in different lengths and are usually made of whalebone.

The 'string' is either horsehair that can be purchased in hanks, or cotton thread coated with beeswax.

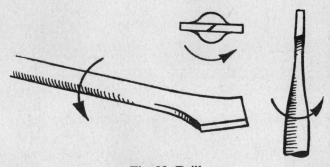

Fig. 20. Drill.

There is a tendency for cotton thread to fray and because of this horsehair is to be preferred.

The tension on the 'string' must allow slip to take place when looped around the ferrule in case the graver should dig into the work. This acts as a safety device and saves the work from damage.

Drilling. As with the turning, there are two methods of drilling in use by watchmakers. Again, one method is to use a lathe and the other is by bow and ferrule.

To use a lathe is to ensure precision. This cannot be said of the bow and ferrule method.

The type of drill used by watchmakers is shown in figure 20, and it will be seen that cutting takes place in one direction only.

The size of the bow will depend on the size of the drill used. For the smallest drills a 9 in. bow with horsehair is sufficient.

The selected drill is mounted in a stock and the lock screw tightened (Fig. 21). A ferrule is secured to the other end of the stock to take the bow. The horsehair is so looped around the ferrule as to rotate the drill in a clockwise direction during a downward stroke.

The stock centre at the ferrule end is positioned in a shallow hole drilled in a steel plate held in the vice.

The work is held against the drill and cutting takes place at each downward stroke of the bow.

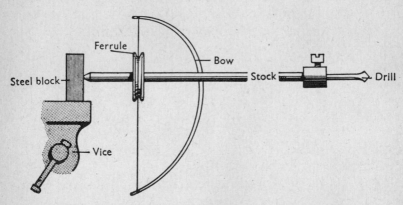

Fig. 21. Drilling.

Part Two

WATCHES

CHAPTER FOUR

The movement

A WATCH movement of high quality is necessarily complicated. Many hundreds of parts are used, each one cut and fashioned by craftsmen of unquestionable skill.

Some of these parts are so small they have to be seen under a powerful magnifying glass before their perfection of design and manufacture can be fully appreciated.

The term 'movement' applies to the complete watch less the case and it is measured in lignes.

The ligne was originally a French measure, being $\frac{1}{12}$th of a pouce or French inch. It travelled to England with the French watchmakers. The ligne was further divided into 12 douzièmes.

The ligne used today by the watch industries in Europe is the French measure—1 ligne is equal to 2·55883 millimetres. This is approximately $\frac{3}{32}$nds of an inch.

The American watch industry uses a measure originated by Aaron L. Dennison in 1850 employing a series of numbers with zero size as a basis. This is equal to $\frac{35}{32}$ths of an inch. Movements larger than 0 are identified by full integers such as 1, 2, 3, etc. Movements smaller than 0 are identified as 1/0, 2/0, 3/0, 4/0, etc. Each step is equal to $\frac{1}{30}$th of an inch.

There is a further complication that jewellers also use lines, but the jewellers' line is $\frac{1}{40}$th of an inch.

The use of the ligne or line as a measure is not consistent with modern manufacturing methods and a number of watch manufacturers have changed over completely to millimetres.

29

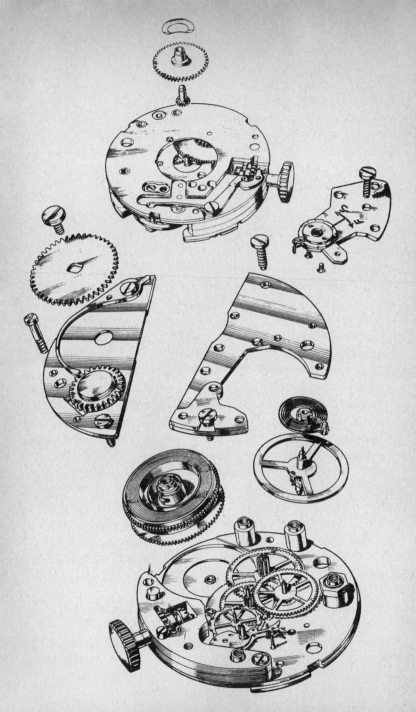

Fig. 22. Exploded view of general construction of inexpensive men's wrist-watch movement.

Figure 22 illustrates an inexpensive men's wristwatch fitted with a pin pallet escapement.

A watch movement consists of a train of wheels with power at one end to drive them, and a means of controlling their speed at the other end (Fig. 23). To this is added the motion work and hands to register the wheel train speed on a dial.

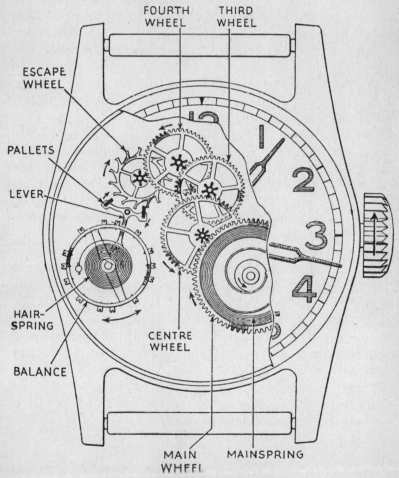

Fig. 23. Wheel train and balance.

It is of interest to note that most wheel or gear trains in common use are designed to reduce speed and increase power. In mechanically powered timepieces the reverse is the case.

Motive power. Power is provided by a mainspring coiled in a barrel. The inner end of the spring has a rectangular hole known as the eye. This is hooked to an arbor in the centre of the barrel. The outer end of the spring is hooked to the inner face of the barrel side. The mainspring occupies about one-third of the space inside the barrel. The action of winding a watch spring causes the arbor to rotate and the inner end of the mainspring to wind itself round the arbor.

There are three methods of using the mainspring to drive the wheel train and they are known as the going barrel, the stationary barrel and the fusee.

The going barrel. Gear teeth are cut on the outer edge of the barrel and this is known as the main wheel. The arbor is squared at one end to receive a key or a ratchet wheel. During the process of winding the spring, the spring tension is held by a ratchet known as the clickwork.

When the movement is functioning, the arbor remains stationary and the barrel rotates around it.

The stationary barrel. This method will be found mostly in American watches. The barrel remains stationary and the arbor revolves in the centre. At one end of the arbor is fitted a toothed wheel.

The fusee. The time-keeping of old watches was affected by the drop in power when the mainspring ran down, and so a means of compensation was necessary. This compensation was accomplished by the introduction of the fusee.

A cone-shaped pulley was positioned next to the barrel. The pulley had a spiral groove machined in its face in which a chain was placed. One end of the chain was hooked to the base of the cone and the other end of the chain was hooked to the outer face of the barrel side.

Gear teeth were cut on the bottom edge of the cone and formed the main wheel. When the mainspring was wound, the chain was pulling on the smallest diameter of the cone with minimum leverage.

As the spring unwound itself and the power became progressively less the chain was being transferred from the fusee and was winding itself on to the barrel. This caused the chain to pull on an ever increasing diameter of the cone with a proportionate increase of leverage. By this method suitable compensation was achieved.

The wheel train. The teeth of the main wheel mesh with the leaves of the centre wheel pinion. In the same way the centre wheel meshes with the third wheel pinion, the third wheel meshes with the fourth wheel pinion, and the fourth wheel meshes with the escape wheel pinion.

The centre wheel rotates once every hour and therefore carries the minute hand. The fourth wheel rotates once every minute and so carries the seconds hand.

The escapement. Some means of controlling the speed of the wheel train is now required. This is done by the escapement which consists of an escape wheel, a lever and a balance.

The balance is to a watch as a pendulum is to a clock. The balance wheel is mounted on a staff that has a very fine pivot at each end both of which operate in jewel pivot holes.

The fitting of a spring to the wheel provides self-contained motive power enough to cause the wheel to vibrate many times before coming to rest. The more accurate the balance and the lower the frictional resistance at the pivots, the greater will be the number of vibrations performed by the balance.

The escape wheel is controlled by the balance through the lever. One end of the lever is fitted with two pallets which allow the escape wheel to revolve one tooth at a time.

When a tooth of the escape wheel comes into contact with one of the pallets, that end of the lever is pushed sideways which causes the other end to transmit a small impulse to the balance. It is this small impulse caused each time the escape wheel moves that prevents the balance from slowing down and stopping.

It will be seen therefore that the wheel train moves in a series of jumps.

To ensure sustained accuracy of time-keeping, each part of the movement must be machined with precision, all pivots finely polished, all friction surfaces adequately lubricated, and the movement must be enclosed in a dust and moisture-proof case.

Everything possible must be done to reduce friction to a minimum. So-called jewels are fitted as pivot bearings. These jewels derive their name from the time when watchmakers used rubies. This practice no longer applies, instead a synthetic material of equal hardness is used.

The number of jewels used in a watch varies with the quality of the movement. The usual number is 7, 15, 17 or 21.

The seven-jewelled movement indicates that only the escapement is fitted with jewels. The majority of jewelled watches have fifteen jewels which takes them up to the third wheel.

In addition to jewels being used as bearings for wheel pivots, they are also used as caps for the balance wheel pivot bearings, these are known as end-stones.

After a movement has been oiled, jewels and end-stones retain the oil by capillary action.

The train of wheels and the escapement are supported by two plates known as the bottom plate which is beneath the dial, and the top plate. Sometimes the top plate is dispensed with and a number of bridge pieces or cocks are used instead.

Between the bottom plate and the dial is the motion work. This consists of a cannon pinion, an hour wheel and a minute wheel.

Winding mechanisms. There are three types of winding mechanisms. Early watches were supplied with separate keys, and were wound through a hole in the back.

Modern watches are keyless, the winding mechanism being part of the movement.

Lastly there is the automatic type. The action of moving one's wrist causes a weight to swing which in turn winds the mainspring. Overwinding is prevented by a slipping clutch mechanism.

CHAPTER FIVE

Overhauling and cleaning

IF a watch develops a fault then it must be traced and put right.

If the movement is clean and oiled there is usually no need to carry out any additional work other than check the watch for time-keeping.

In this chapter we are going to discuss generally the complete overhaul and cleaning of wristwatch movements. The remaining chapters in this section deal with specific assemblies of parts in greater detail and must therefore be read in conjunction with this chapter.

Plates I to VI are those of a $5\frac{1}{4} \times 8\frac{3}{4}$ ligne 15-jewel lever movement. They serve to show the sequence of assembly of the parts in a modern wristwatch and the identification of parts by name.

Too much emphasis cannot be placed on the need for careful inspection as the work of dismantling continues. A great deal of conscientious work can be put into the inspection and cleaning of individual parts after dismantling but when the movement is once again assembled the original fault may still be there. If the history of the watch is unknown to you then it must be borne in mind that some of the parts may not be the originals and that incorrect replacements may have been fitted or that bad fitting may have taken place. Before opening the case examine the watch closely. Frequently much useful information can be obtained that might otherwise have remained obscure.

Methods of opening watch cases are dealt with in Chapter 13 and you are advised to read this before proceeding any further. If when the case has been opened a considerable amount of dust is found inside, then the case needs some kind of attention.

Operate the hand set and turn the hands in the normal direction and listen for fouling against the glass or dial. With a keyless watch it should be possible to pull the winding stem out and push it in without undue force or looseness. The feel of the action should be smart and positive. Test the side-shake of the winding stem. There should be just sufficient to prevent binding when being operated. This is very important because a loose fitting winder is an open invitation to dirt and dust to enter the movement. Examine the case for signs of it having been dropped or knocked. A blow on the case frequently results in broken balance pivots if shock absorbers are not fitted. Remove the glass and inspect it for signs of fouling by the hands. Carry out the same inspection on the dial. The most likely fault will be fouling by the tip of the minute hand. Lightly touch the hands and apply side pressure to see if they are loose.

Operate the hand-set and hold the watch so that it can be viewed from the side, turn the hands and observe their movement. This will disclose any tendency to foul the dial and to foul themselves. When carrying out this check make sure that the dial is flat with the bottom plate and is firmly secured.

With the aid of an eye-glass inspect the seconds hand to see if it is free of the dial face and that there is clearance between the seconds hand pipe and the hole in the dial.

If when turning the hands it is noticed that the tip of the minute hand rises and falls, this is an indication that the centre wheel is not upright. But if the minute hand rotates at a constant height above the dial and the hour hand rises and falls, then the centre arbor can be suspected of being bent.

The centre arbor can easily be checked by spinning it between a pair of callipers. If the arbor is bent, lay it on a flat steel stake and lightly tap it with the peaning end of the watchmaker's hammer.

To correct the uprightness of a centre wheel necessitates one of the centre holes being bushed. This calls for the use of equipment that is unlikely to be in the possession of a beginner, e.g. a lathe and a punch and stake. In this event the work of bushing will have to be sent away.

We now check the hour wheel to make sure that it has some

shake. With a pair of tweezers hold the hour hand by its socket and lift the hour wheel up and down. If there is no end-shake remove the minute hand and try again. If end-shake is now present it indicates that the hour hand or the hour wheel pipe is fouling the minute hand and that it is necessary to lower the hour hand on the hour wheel or reduce the length of the hour wheel pipe.

If, however, removal of the minute hand makes no difference look to see if the hour wheel pipe is fouling the dial hole or if the hour wheel is binding on the cannon pinion.

The latter two faults will also eliminate side-shake which is the next check to make. Additionally, if the depth of mesh between the hour wheel and the minute wheel pinion is excessive the side-shake will be reduced or eliminated altogether. It is essential that the hour hand has complete freedom of movement in both directions otherwise the resistance offered up will have a retarding effect on the time-keeping of the watch if not stopping it completely.

Remove the minute hand and the hour hand with the levers illustrated in figure 10, and at the same time take precautions against damaging the dial by placing beneath each lever a piece of folded tissue paper. Hold the levers, one each side of the hand to be removed, and by applying light equal pressure downwards the hand will come away. The seconds hand is removed by the action of lifting off the dial.

The movement may now be taken from its case. There are a number of methods used to secure movements so let us consider those that we are most likely to meet.

The movement of a keyless watch fitted to a two-piece case, such as is illustrated in figure 69, is made to slide into the case and is held in position by the bezel. To remove it pull out the winder into the set-hand position and by holding the winder gently ease the movement from the case. The use of a small screwdriver as a lever will assist this operation if the movement sticks and fails to lift out squarely.

The keyless watch with the three-piece case is rather different.

The movement, less the winding stem, is put into the case middle from the front and is prevented from passing right through by the aperture of the case middle being smaller at the back.

The movement is held in position by dog screws, usually two in number. These screws have large diameter heads and are inserted at the back of the movement. In screwing them into the watch plate the heads tighten on to a lip of the case middle and hold the movement secure.

Some dog screws have almost half of their heads cut away. These screws only have to be turned slightly to bring the flats of their heads in line with the case lip and the movement is freed without the necessity of removing the dog screws.

With these watches the winding stem passes through the case middle before entering the movement and is therefore inserted after the movement is in position.

To take the movement out, slacken the pull-out piece setting bolt, withdraw the winding stem, slacken or remove the dog screws and carefully push the movement forward and out.

Some of the earlier key-wound watches have their movements hinged to the case middle with a spring catch diametrically opposite to hold it. To take the movement out necessitates the removal of the hinge pin.

Whatever the method used, however, careful examination will indicate the procedure for dismantling.

Having taken out the movement we now have to lift off the dial. In the majority of watches the backs of the dials have two copper feet soldered diametrically opposite each other. These feet pass into holes in the bottom plate and are secured by screws entering the bottom plate from the side. Slackening of these screws frees the dial. Usually a little assistance is required because the screw tension causes slight bending of the feet preventing them from being lifted straight out. Insert a thin blade of a pocket knife under the dial close to the feet and slightly twist the blade so that the pressure on the dial takes place at the base of the feet. This method will prevent the dial from bending, a precaution which is essential with enamel dials in particular to prevent cracking of the enamel. When lifting the dial off take care not to lose the seconds hand if there is one fitted.

Another method used to secure the dial is the fitting of dog screws.

These screws have a conical flange half way down their length. One side of the flange is cut away. The screw is inserted in the plate parallel and adjacent to the dial feet and turned so that the edge of the flange passes into a slot cut in each dial foot. By rotating the screws half a turn inward the cut away portion faces the dial feet and enables the dial to be withdrawn.

The dials of some of the cheaper watches are made with small tongues on the edge that are bent over to grip the movement when the dial is in position. When these tongues have been bent a few times they snap off and means of securing the dial are lost. In handling a dial of this kind it is essential that the tongues are bent as little as possible.

Having removed the dial inspect the underside for signs of fouling. The dial may have been pressing on the minute wheel pinion. If this happens the timekeeping will be erratic. The minute wheel pinion may bind only in certain positions. When the watch is not being worn the pressure of the dial on the pinion may just be sufficient to stop the movement, but the action of disturbing the watch may cause the movement to function again. During the time the watch is worn the effect may be a slowing down at intermittent periods.

Should there be some doubt as to whether fouling is taking place, a drop of oil on the pinion can be used as an indicator.

Replace the dial, operate the hand-set and rotate the motion work. Remove the dial and inspect again. Any trace of oil on the underside is an indication that there is insufficient clearance between the dial and the minute wheel pinion.

There are two methods of increasing this clearance; by removing metal from the underside of the dial or by reducing the height of the minute wheel pinion.

In the case of a metal dial take a chamois leather and fold it into four thicknesses and place it on a flat block of wood on the bench. Over the chamois leather place two thicknesses of tissue paper, then lay the dial face downward on it.

Scraping is done with the tip of a very sharp knife. The strokes must be as light as possible to prevent bruising of the dial face. Continue scraping until all the pinion marks have been removed and then fit the dial to the movement and test again by the oil method.

To reduce the thickness of an enamel dial a carborundum stick is required. These are made specially for the job and can be obtained from your supplier.

Support the dial as for a metal dial. Moisten the carborundum stick in water and rub it fairly heavily over the affected area in a circular motion. Continue this grinding action until all traces of fouling have been removed or until the copper dial shows through. If the copper does show before the pinion marks have been removed then it is advisable to stop work on the dial and have the height of the minute wheel pinion reduced in a lathe.

Earlier in the chapter it was mentioned that before removing the hands, a little side pressure should be given to them to find out if they were tight on the wheels. In the case of the minute hand although the hand may have been tight on the cannon pinion, nevertheless the cannon pinion may have been loose on the centre arbor.

This is a very common fault with the snap-on type of cannon pinion and is a constant source of trouble.

The centre arbor is machined with a groove all round it. The cannon pinion is similarly machined and the base of the groove is then burnished inward causing an internal bulge. It is this bulge that locates the pinion on the arbor and provides the friction drive.

The remedy against slipping is to reburnish the cannon pinion groove until it grips the arbor once again.

Lift off the minute wheel and check the tightness of the minute wheel post. Then hold the minute wheel between finger and thumb and grip the minute wheel pinion by inserting the points of the tweezers between the leaves. Rock the tweezers to find out if the minute wheel is loose on the pinion. Replace the minute wheel on its post.

Depthing of the wheel teeth must now be tried. First remove the hour wheel and try the depthing between the cannon pinion and the minute wheel. Then replace the hour wheel and try the depthing between the hour wheel and the minute wheel pinion. After carrying out these checks the wheels must be rotated so that depthing can be tried with a different set of teeth. Repeat this until each wheel and pinion has been tried in four different positions.

On the bench in front of you place a sheet of stiff white paper on which the parts can be placed as they are removed from the movement. The parts should be kept together in groups complete with any screws. Sometimes the screws in one group are similar but one is slightly longer than the others. It is important that a screw is not used where a shorter one should be. There is the possibility of the end of the thread fouling something which may prevent the screw from being tightened or may cause the watch not to function.

The screws should, therefore, be placed in or near to their respective holes in the part to which they belong.

Now turn the movement over in readiness to remove the balance.

With the mainspring partly wound carry out an inspection of the escapement and make notes of any faults. Details of this inspection are contained in Chapter 10.

If the watch is fitted with a cylinder escapement the mainspring must be let down before the balance is removed, otherwise the train of wheels will be driven by the mainspring at high speed which will invariably lead to damage of the pivots.

To let the mainspring down hold the winding button, or the key in the case of a key-wound watch, and make a slight turn in the direction of winding. This will partly raise the click which can then be fully released from the ratchet wheel by the point of a pegwood stick.

The tension of the mainspring is now directly transmitted to the winding button which, if allowed to rotate slowly between the finger and thumb, will permit the mainspring to unwind.

Care must be taken not to let go of the button whilst the click is held back because the mainspring will then unwind itself at high speed and cause damage to its centre. If it is felt that finger control of the button is being lost, release the click into the ratchet wheel and try again.

When the movement is in this position it must not be laid on the bench and worked on because the centre arbor that carries the cannon pinion and the fourth wheel pivot that carries the seconds hand will be projecting.

Either the movement rests at an angle on the bench and is supported in the hand, or a watch movement stand must be used. Adjustable stands can be purchased but a length of metal tube about $1\frac{1}{2}$ in. long, squared at the ends and of suitable diameter will do quite well.

Take out the screw holding the balance cock and very carefully lever the balance cock up. A small notch will be seen on the edge of the cock into which the blade of a small screwdriver can be inserted. Because the leverage will be concentrated at one end there will be a tendency for the cock to press down on the balance pivot at the other end. This must be prevented and once the balance cock has been moved it must be brought level again by raising the other end in tweezers.

Once the balance cock pins have been withdrawn from the bottom plate the cock can be lifted from the movement with the balance wheel and spring hanging from it.

You may find there is a tendency for the balance wheel to catch and not come away. This will be because the roller is caught in the lever fork. If this happens any further attempt to lift the balance will only result in the coils of the hairspring being pulled out.

To free the roller, the cock is held stationary above the movement, and the movement is slowly turned until the roller is freed.

Lower the balance to the bench and when the bottom pivot

touches the paper roll the cock over on its back. In doing this the spring will cause the balance wheel to do the same. The balance wheel is now held in the tweezers and positioned so that the balance pivot rests in the jewel hole in the regulator index.

The balance now has to be separated from the cock. First, examine the index pins. Some watches are fitted with one fixed pin and one movable pin. The movable pin has a lip on the top that bridges the gap between the two pins and so prevents the outer coil of the hairspring from coming out. This movable pin has a screwdriver slot in the end enabling the pin to be turned thus swinging the lip clear and making it possible for the hairspring to be removed.

Now remove the stud from the balance cock. Studs are either held in by a screw or they are a push fit. If the stud is the latter type, grip it firmly with a pair of strong tweezers and gently ease the stud out, taking care not to alter the bend or set of the hairspring.

Lift the balance clear and place it on the white paper along with the balance cock screw.

It is advisable at this stage to leave the index assembled to the cock and remove it only when you are ready to clean it.

If the movement is not fitted with a cylinder escapement then you will not have unwound the mainspring.

The pallets are next to be removed and so we must let the mainspring down.

Having done this, remove the screws from the pallet cock, lift the pallet cock away and remove the pallets.

If the wheel train starts to rotate, it is only because by releasing the click from the ratchet wheel it is not possible to completely let down the mainspring. However, what little energy remains in the spring quickly spends itself with no damage to any part.

Before removing the wheels, turn the watch over so that the cannon pinion can be removed. Because the cannon pinion is a friction tight fit on the centre arbor all that is required to remove it is a straight pull. Care must be taken, however, in avoiding damage to the pinion when it is being held. One of the safest and most simple methods is to grip the cannon pinion in a pair of brass-faced pliers.

With the cannon pinion removed turn the movement over again.

Before proceeding with the dismantling of the wheel train we must test the depthing. Take a pegwood stick and shave it to a blunt point. Press the end of the stick on to the fourth wheel pivot and hold the wheel. With another pointed pegwood stick touch the third wheel and try the shake of the third wheel teeth in the fourth wheel pinion. Repeat this process for the other wheels.

In Plate IV it will be seen that the escape wheel, fourth wheel, third wheel and the centre wheel are all held in position by a train wheel bridge. This arrangement does not apply to all watches. In some movements the centre wheel is held in by a top plate and the remaining wheels are supported in the same cock or bar. Then there is the bar movement in which each wheel has its own bar. This method is preferred by many watch repairers but the other methods have obvious advantages.

Remove the screws from the wheel train bridge and carefully lift the bridge off. Some bridges will have small notches similar to that previously mentioned in the balance cock for the insertion of a lever. The bridge or cock has locating pins underneath and as soon as they are free of the bottom plate the bridge will come away. Lift out the wheels and place them with the bridge and the screws.

Now remove the click, click spring, transmission wheel and ratchet wheel from the barrel bridge. Some watches are fitted with transmission wheel screws with a left-hand thread and this must be borne in mind when attempting to loosen the screw. It is as well to get into the habit of trying to turn the screw first one way and then the other slowly increasing the pressure until the screw eventually turns.

Remove the barrel bridge and lift out the mainspring barrel. This leaves the bottom plate with the balance endpiece and the keyless work.

Lift the crown wheel and the castle wheel from the plate.

Remove the pull-out piece check-spring, the pull-out piece screw and the pull-out piece itself. Lastly remove the winding stem.

Remove the balance endpiece screw and allow the endpiece to drop out. If the oil keeps the endpiece in position take a pointed pegwood stick and turn the endpiece. Now with the same stick the endpiece can be pushed out from the other side of the bottom plate by inserting the stick in the endpiece screw hole.

The dismantling of the watch is now complete with the exception of the balance cock index and the mainspring barrel.

It is the intention that these two assemblies remain as they are until we are ready to clean them.

On the bench in front of us we now have all the parts on the sheet of white paper. The parts have been placed together in their groups and we have made quite sure that the screws have been placed in or alongside their screw holes.

This grouping of the parts will considerably ease the task of assembly as you will see when you first tackle the job.

We are now ready to start the cleaning and so we will commence by getting together our equipment and materials.

The parts are washed in a cleaning fluid (see Chapter 2). A small clean pan will be required for use as a washing bath. A pan measuring 3 in. × 2 in. × ¾ in. will be sufficient.

Two watchmaker's brushes will be required, one with medium bristles and the other with soft bristles, and a block of billiard chalk to charge the brushes with.

We will want the pegwood sticks, some pith, tissue paper, and a clean soft linen cloth.

Lastly we require some means of keeping the parts perfectly clean after we have finished with them. Small glass dome-shaped covers can be purchased for this purpose but alternatively a good method is to obtain a clean shallow pan, large enough to contain all the parts in their groups, and cut a piece of white paper to fit in the bottom. Then we will want a piece of clear glass to place over the pan as a dust cover.

Now that our tools and materials are laid out on the bench we can commence the process of cleaning.

Pour some cleaning fluid into the washing tray and replace the lid of the container. Take both cleaning brushes and charge them with chalk by stroking the billiard chalk a few times with the bristles. Make sure the full length of the brushes is treated.

With the exception of the dial, the hands, the pallets of a lever escapement and the mainspring barrel assembly, all the parts are immersed in the cleaning fluid. By holding the parts with tweezers and moving them about dirt and old oil will be washed off.

The parts are then lifted out and placed on tissue paper to drain. In a warm room the cleaning fluid will soon evaporate leaving the parts clean and dry.

Each part is then polished by brushing to ensure absolute cleanliness. Bearing holes are cleaned with a pegwood stick shaved to a fine point, and pivots are cleaned by rotating them in a pith stick.

Once the parts have been removed from the cleaning fluid they must not be touched by hand. If it is necessary to hold them in the hand do so with tissue paper, otherwise hold them to the bench with tweezers when working on them. This is important because finger-marks can cause unsightly stains.

After each part has been cleaned and polished it must be examined carefully through an eye-glass for signs of dirt, corrosion and damage and if all is well it can be placed under cover in the tray for cleaned parts.

Note the position of the index and then dismantle the balance cock. Place it on its back on the bench and hold it down by the tip of a finger whilst removing the two small screws. In some movements these screws are very small and great care must be exercised in their handling to prevent loss.

If the balance is fitted with shock-absorbers then reference must be made to Chapter 12 for instructions on dismantling.

When the balance cock has been taken to pieces the parts are transferred by tweezers into the washing bath. The parts are left to soak and in the meantime the balance is dismantled.

Make a note of the relative position of the hairspring collet to the balance wheel so that it can be replaced on the staff in the same position. Hold the balance wheel between the tips of the first and second fingers and the ball of the thumb. Insert the blade of an oiler or a very small screwdriver into the slot of the collet. Slight pressure should widen the slot and permit the collet to be withdrawn from the balance staff. Place the balance wheel and the hairspring into the washing bath.

The parts of the balance cock can now be lifted out of the bath. Place each part on a piece of tissue paper and allow the cleaning fluid to drain off and be soaked up by the tissue. The cock can be dried in the linen cloth.

Hold the balance cock in tissue paper and brush gently with the medium brush. Apply the brush in straight strokes and keep the grain flowing with the length of the cock.

The remaining parts are laid on the sheet of white paper and are held by the tweezers whilst being brushed. Use the soft brush for these parts.

The jewel bearing in the balance cock must be cleaned and to do this a piece of pegwood is shaved to a fine point, inserted into the bearing and is given a turn or two applying very little pressure. Remove the pegwood stick, shave off the dirt and insert it again, this time from the other side.

Assemble the balance cock and replace the index in its original position.

Now lift out the balance wheel and allow it to drain off on the tissue paper. Brush the wheel carefully with the medium brush taking great care to avoid knocking the pivots. In some movements the pivots are so fine the slightest knock will result in breakage.

Finish off by polishing the pivots with a pith stick.

Place the hairspring on tissue paper and lay another piece of tissue paper on top. Continue to renew the paper until all cleaning fluid has been absorbed. Then lightly dab between the coils with the tip of the soft brush.

Examine the balance wheel, roller or cylinder, and hairspring very carefully and then replace the hairspring collet on the balance staff in its original position.

Lift out the pallet cock and screws and clean them in the same way. The pallets (if from a lever escapement) have not been immersed in the cleaning fluid. This is because some shellacs are not impervious to gasoline and it is therefore considered better not to soak them but to immerse them in the fluid, agitate them to remove the dirt and lift them straight out again.

Place them on the tissue paper to dry. Hold them with the tweezers and thoroughly brush them with the soft brush. Finish off by polishing the pallet stones with a pith stick.

Clean the leaves of the escape wheel pinion with a pointed pegwood stick. Rub the stick up and down until the leaves are bright. Well brush the wheel with the soft brush and then clean each tooth and both pivots with the pith stick.

The teeth of an escape wheel fitted to a pin-pallet escapement call for special attention. The constant action of the pallet pins working in the escape wheel teeth causes a build-up of dirt at the base of the teeth. This dirt is forced into the corners by the pins and becomes quite hard.

Normal soaking and brushing is insufficient to remove the dirt and unless it is removed the shake of the roller will be affected.

If a sharp knife blade is placed at the root of each tooth and lightly drawn across, the dirt will come away. Handle the blade with care because careless handling can result in the removal of metal which would have a serious effect on the escapement. Finish off with a stiff brush.

Proceed with the rest of the movement in the same way making sure each part is examined under the eye-glass when it is finished. Remember that the large pieces such as cocks, bridges and plates can be dried in the linen cloth.

The barrel is next to receive attention. Before removing the barrel cover make a note of its position in relation to the barrel. It is essential that the cover be replaced in the same position.

Insert the blade of a screwdriver in the slot of the barrel cover and gently ease the cover off.

Before proceeding any further a close inspection of the main-spring and barrel is necessary. If the cover of the barrel is a good fit and the condition of the mainspring is clean and well oiled, and no dirt is present, it is unnecessary to remove the spring.

To take a mainspring from its barrel and allow it to fully expand does more harm than all the good that cleaning can do. If the spring looks clean and healthy, leave well alone.

If, however, it is decided that the spring must be removed, then proceed as follows.

Grip the arbor in the pin-vice and with a slight clockwise turn to free the hook, lift the arbor out.

The removal of the spring must be done with care otherwise it may be rendered unfit for further use.

Hold the barrel in the left hand and a pair of strong tweezers in the right hand. Grip the centre coil of the spring and gently ease it out.

When the spring shows signs of moving without assistance hold it back with the ball of the left thumb and slowly release the pressure.

By this means the spring will unwind itself from the barrel but at a controlled rate.

If the centre was to be pulled right out and the spring allowed to fly open it would adopt a conical shape which would render it unsuitable for further use.

In any case, a spring that is let down rapidly after being under tension for a considerable period invariably breaks.

Wash the parts in the cleaning fluid, brush the barrel and polish the arbor with the pith stick. The spring is wiped with tissue paper.

Take a piece of tissue paper and fold it over. Place it around the outer coil and with a pair of tweezers slide the paper along the coils to the centre. Do not open the spring but allow it to retain its natural curvature.

There now only remains the dial to be cleaned. There is very little that can be done however.

If the dial is enamel and has smeary marks on its face it may be wiped with a soft cloth moistened with the cleaning fluid and then lightly polished with a dry piece of the same cloth. Nothing can be done with a crack but if dirt has found its way into a crack showing it up as a thin black line the dirt can be removed by brushing along the crack under running water.

Any attempt to clean a silver or gilt dial almost always results in the dial looking worse. It is safer to leave these alone. A gold dial can be wiped with a moist chamois leather.

This completes the cleaning of the movement and the next step is to assemble the parts. The oiling will be done progressively during the work of assembling the movement.

We will start with the mainspring and barrel. The correct method of fitting the spring into its barrel is by using a mainspring winder. There are a number of ways in which a winder can be used.

The eye of the spring is attached to the hook of the winder and the spring is wound in the hand, a firm hold being maintained by finger pressure to prevent it unwinding. The winder is unhooked from the spring and by careful manipulation the spring is transferred to the barrel.

Another method is to attach the spring to the winder as before. Hold the barrel in the left hand against the spring and in a position ready to receive the spring when wound. Commence to wind holding the spring with the fingers and when the diameter of the spring is less than that of the barrel, the barrel will move forward over the spring.

The type of winder commonly used in the United States employs a set of dummy barrels with plungers. A dummy barrel, smaller in diameter than the barrel of the watch, is selected. The spring is wound into the dummy and then transferred to the watch barrel by pushing it out with the plunger.

It does not follow that without a mainspring winder it is not possible to fit the mainspring. By placing the arbor in the barrel and gripping it behind the barrel with a pin-vice, the eye of the mainspring can be hooked to the arbor and can then be wound in. This method, although practised, is not recommended because the angle at which the spring is fed into the barrel tends to create distortion. Due to the very small clearances between a mainspring and the barrel cover it will be appreciated that an almost imperceptible amount of distortion towards being conically shaped can result in the edge of the mainspring fouling the cover. This would cause the spring to stick and have a jerky action.

There is yet another method of fitting a mainspring, and it is by hand. Again, the method is bad practice for the same reason given above. The centre of the spring is positioned in the barrel. The barrel is held between the first finger and thumb of each hand and rotated and at the same time the spring is fed in by the ball of the thumb.

It has been established that one of the main causes of mainspring breakage is rust. A very tiny speck will weaken the spring and under tension the spring is liable to break at that point.

Rust can result from handling and therefore it is important that the mainspring be touched by hand as little as possible. Place the arbor in position and make sure the eye of the spring is over the hook of the arbor. The centre coil should now be wrapped round the arbor. If it is not, remove the arbor and close the centre of the spring slightly. When the centre coil is correctly set it is necessary to push it to one side to get the arbor in.

Charge the watch oiler with oil and touch it three times around the edges of the spring.

Let us pause a moment here because this is the first time any reference has been made to the application of oil, and we must therefore have an understanding of the subject before we can proceed any further.

There are four things to consider; where to oil, choice of oil, quantity to apply, and lastly how to apply it.

Taking them in that order then we can say that oil is required wherever two or more surfaces are rubbing together. Ideal lubrication can be better understood if we imagine two perfectly flat surfaces, one on top of the other, completely separated by a thin film of oil. If a means of retaining that film of oil can be provided, then the top plate can be moved over the bottom plate with almost no resistance.

Below is a table listing the points that require lubrication in a wristlet watch such as the one illustrated. This will serve as a guide for other watches.

Top Grade Watch Oil	*Top Grade Clock Oil*
Escape wheel pivot holes Fourth wheel pivot holes Third wheel pivot holes Centre wheel pivot holes Three teeth of the escape wheel Balance pivot holes	Crown wheel sleeve Cannon pinion friction groove on centre wheel Pull-out piece pin Winding stem pivot Pull-out piece Return bar stud Return arm Barrel arbor pivot holes in plate, bridge, barrel and cover Mainspring

The second consideration is also dealt with in the table in that the choice of oil is given.

As to the quantity, this, of course, cannot be stated in degrees of measurement. One drop of watch oil and one drop of clock oil from a large watch screwdriver is sufficient to lubricate two or more watches as illustrated.

Too much emphasis cannot be placed on the need to guard against over-oiling. When one bears in mind it is only the areas of frictional resistance that need oiling and then consider how very small some of these areas are, it will be appreciated that the correct amount of oil to apply in these cases would be difficult to see without the aid of an eye-glass.

Excess oil creeps on to watch parts causing minute particles of dust and fluff to adhere. This accelerates the rate of drying up the oil. Parts of the watch that it is essential should be free of oil become contaminated, such as the escapement banking pins, the ruby pin, the hairspring and so on.

The aim in cleaning and oiling a movement is to finish with a meticulously clean and perfectly dry watch with just sufficient oil applied in exactly the right spots.

If two lubricated surfaces, one flat and the other curved, are placed together the oil will concentrate around the apex of the curve. This attraction is the capillary action of a fluid and is the reason for an oiler being shaped as it is. When the oiler is charged with oil, the oil remains on the end and does not run up the blade. Then when the tip of the oiler is brought into contact with the part to be lubricated, the capillary action will cause the oil to be transferred.

To oil a watch movement two oil cups are needed, one for watch oil and the other for clock oil. Dip the large watch screwdriver into the bottle of watch oil and transfer one drop to the cup for watch oil. Charge the clock oil cup in the same way. Wash the screwdriver in the cleaning bath before using it again.

Make sure that the oiler is free from oil before touching it to the oil in the cups. This will enable you to control the amount of oil picked up. If the oiler blade is moist with oil with perhaps a smear along the blade, then the effect of touching the oiler to the oil is to pick up an excess amount.

If this should happen, no control can be exercised over the amount of oil deposited by the oiler when applying it to the movement because the capillary action will attract almost all of the oil and deposit it on the watch.

Now that we have a better knowledge of the technique of oiling we can continue assembling the movement.

Now that the mainspring has been oiled the barrel cover may be snapped on. Covers are made to fit in one particular position and it is important that the cover is always replaced in this position.

Many are marked with dots so that when the dot on the cover is opposite the dot on the barrel the two are in the correct position in relation to each other.

Make sure the cover fits square and flush. Hold the arbor in the brass-faced pliers and turn it in the direction of winding until you feel the hook on the outer end of the spring find its anchorage on the barrel wall. Apply a touch of oil to each end of the arbor where it operates in the barrel.

Place the bottom plate on the movement stand. Insert the escape wheel followed by the fourth and third wheels. Oil the lower centre pivot and insert the centre wheel. Fit the train wheel bridge taking care the recess does not catch the third wheel as the bridge is lowered.

When fitting the train wheel bridge place a piece of tissue paper on it and press down very lightly with the finger. The top pivots of the wheels can be guided into their bearings with a pair of fine tweezers and the bridge will then lower into position without any forcing.

Oil the pull-out piece screw and fit it in position. Insert the barrel and refit the barrel bridge. Oil the barrel arbor in the bottom plate and in the barrel bridge.

Lubricate the seating of the transmission wheel and replace the transmission wheel and the barrel ratchet. Don't forget that the transmission wheel screw will probably have a left-hand thread.

Now turn the movement over. Oil the centre arbor and fit the cannon pinion. Insert the crown wheel and the castle wheel. Oil the winding stem and fit it in position. Fit the pull-out piece.

Oil the ratchet teeth and the castle wheel groove. Oil the return bar stud and replace the return bar and the spring. Lightly oil the keyless work.

Check the end-shake and the side-shake of the wheel train, and oil the top and bottom pivots. Wind up the mainspring one click of the ratchet and make sure that the wheel train has complete freedom of movement.

Fit the pallet and the pallet cock and oil the pivots. Wind up the mainspring about four clicks and press the lever lightly on one side with the tweezers. Make sure that the lever runs to the opposite banking pin with a lively action.

The balance now has to be fitted to the balance cock. During the dismantling it was considered inadvisable to unpin the hairspring from the stud and therefore all that is now required is for the stud to be replaced in the balance cock.

Some studs are secured to the cock by a screw whilst others are a push fit. Among those that are held by a screw will be found some that are machined with a straight groove into which the end of the screw is located. This ensures that each time the stud is replaced it always takes up the same position.

With the others it may be necessary to turn them slightly one way or the other to line up the outer coil with the index pins.

Hold the balance cock in the tweezers and allow the balance to hang from it. Lower the balance into the movement and with careful manipulation the balance will find its own position as the cock is lowered.

Place a piece of tissue paper on the cock and very gently press down with the finger making sure that the balance has freedom of movement at all times.

Before the cock is fully down the balance will probably commence to vibrate. Screw the cock down and check the beat of the movement. If the beat is irregular the balance must be removed and separated from the cock and the hairspring collet moved round the balance staff into a new position. This is dealt with in greater detail in Chapter 12.

When the beat is correct remove the balance for the last time. Oil both pivot holes and replace the balance.

Replace the motion work and refit the dial. The movement is now ready for fitting into its case after which the hands can be replaced.

Wind up the mainspring, set the hands to indicate the correct time and put the watch to one side for a check on time-keeping. Any small adjustments that are needed can be made by altering the position of the index.

When the watch is to your satisfaction there is only one more thing to consider and that is whether or not it is necessary to seal the case.

If the watch is to operate in tropical conditions where the humidity is high, or if the wearer is employed in conditions that subject the watch to steam such as a laundry, or if dry dusty conditions exist in a concentrated form, then to seal the case would be a distinct advantage.

Make up a small quantity of a mixture of four parts of beeswax to one part of Vaseline. Smear a thin film of the wax between the joins before closing the case. A small application around the winding stem at the top will complete the treatment.

CHAPTER SIX

Wheel trains

ONE vibration of a balance wheel is the completion of movement in one direction. The total number of vibrations in one hour can be calculated by counting the number of teeth in the wheels and number of leaves in the pinions.

Because the teeth of the main wheel in mesh with the centre wheel pinion are only a means of transmitting power to the train of wheels, timing is calculated from the centre wheel.

Let us assume the train is as shown in figure 24. Because the centre wheel has eight times as many teeth as the pinion of the third wheel it follows that one revolution of the centre wheel causes the third wheel to rotate eight times.

Following on from this, if the third wheel has 75 teeth and the fourth wheel pinion 10 leaves, then one revolution of the third wheel is equal to $7\frac{1}{2}$ revolutions of the fourth wheel.

So far then, we can say that the fourth wheel revolves $8 \times 7\frac{1}{2} = 60$ times to one revolution of the centre wheel.

Finally we come to the escape wheel. The escape wheel pinion has 8 leaves and is driven by the fourth wheel that has 80 teeth. One revolution of the fourth wheel will therefore cause the escape wheel to revolve 10 times.

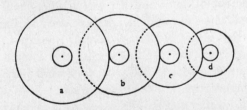

Fig. 24. Wheel train: (a) centre wheel 80 teeth (b) 3rd wheel 75 teeth, pinion 10 leaves (c) 4th wheel 80 teeth, pinion 10 leaves (d) escape wheel 80 teeth, pinion 8 leaves.

Progressive calculations now show that one revolution of the centre wheel is equal to $8 \times 7\frac{1}{2} \times 10 = 600$ revolutions of the escape wheel.

If there are 15 teeth on the escape wheel, and each tooth is equal to two vibrations of the balance wheel, then the total

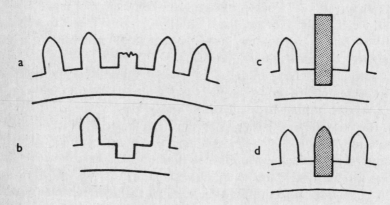

Fig. 25. Fitting new tooth.

number of vibrations to one revolution of the centre wheel is $8 \times 7\frac{1}{2} \times 10 \times 15 \times 2 = 18,000$.

This is more simply expressed as

$$\frac{80 \times 75 \times 80 \times (15 \times 2)}{10 \times 10 \times 8} = 18,000$$

If the number of teeth and leaves on the wheels and pinions are recorded when a watch is dismantled, and the number of vibrations each hour calculated, adjustments for correct timing can be made when the movement has been assembled.

In the example given, 18,000 vibrations were made in one hour = 300 in one minute or 100 in twenty seconds.

Providing a watch that maintains accurate time is available and fitted with a seconds hand, the beats of the watch undergoing overhaul can be counted during a period of twenty seconds and adjustments made as necessary.

Sometimes inspection of the movement reveals a broken wheel tooth.

Broken teeth are not common failures but when they do occur it means a repair or a new wheel. Because of this an attempt at a repair will be good practice and if the completed job is not good enough nothing has been lost; a replacement wheel can still be obtained.

The repair sequence is illustrated in figure 25. The broken tooth is shown at (a). With a thin file cut a slot as at (b). The teeth on one edge of the file can be ground off to assist in filing the corners of the slot square.

A piece of brass pin-wire is selected and two flats are filed opposite each other until the wire fits tightly into the slot as at (c). Cut off the surplus wire and make sure the remainder is parallel to the face of the wheel.

Carefully raise the temperature of the wheel in the repair area by touching it against a soldering iron and then apply some soldering flux. When the wheel is hot enough place a small piece of solder on the base of the new tooth and allow the solder to penetrate through the join to the under-face of the wheel.

Allow to cool off and then file the tooth to shape as at (d).

After the repair make sure that all traces of soldering flux and filings have been washed and brushed away.

CHAPTER SEVEN

Hands, dial and motion work

THE motion work consists of a steel cannon pinion, a brass hour wheel and a brass minute wheel and pinion (Fig. 26).

The cannon pinion has a groove machined round it and the base of the groove is burnished inward causing a bulge. The centre wheel arbor has a corresponding groove cut round it and when the pinion is pushed on to the arbor the internal bulge snaps into the arbor groove and provides a friction tight fit.

The minute wheel and pinion revolve on the minute wheel post which is fixed to the bottom plate.

The cannon pinion drives the minute wheel and the minute wheel pinion drives the hour wheel which revolves freely around the cannon pinion.

The hour hand is pushed on to the end of the hour wheel pipe and the minute hand is fitted over the end of the cannon pinion.

The gear ratios are such that the hour wheel revolves once to every twelve revolutions of the cannon pinion.

By applying the formula in Chapter 6 for calculating gear ratios we find:

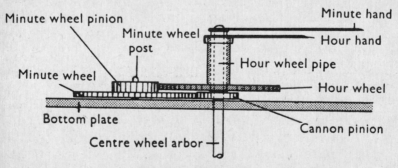

Fig. 26. Motion work.

$$\frac{\text{Minute wheel teeth} \times \text{Hour wheel teeth}}{\text{Minute wheel pinion leaves} \times \text{Cannon pinion leaves}} = 12$$

The most usual wheel trains employed in motion work are shown in the following table.

	Keyless Genevas and Key-wind watches				Modern Keyless English Levers
Cannon Pinion	10	8	10	12	14
Minute Wheel	30	32	40	48	28
Minute Pinion	8	10	12	14	8
Hour Wheel	32	30	36	42	48

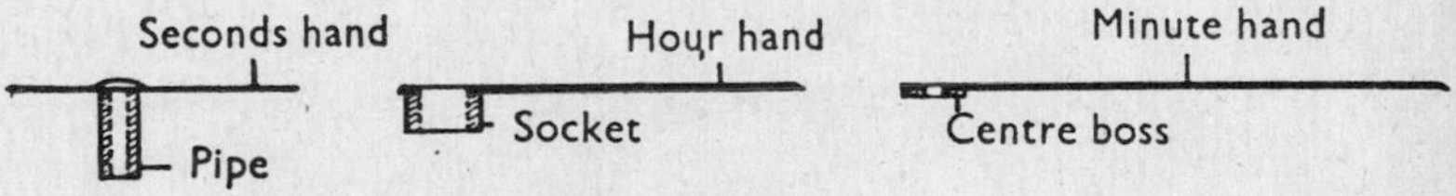

Fig. 27. Hands.

MOTION WORK TRAINS

Hands (Fig. 27). All too often the fitting of hands receives insufficient attention. Badly fitted hands are a common cause of watch failure. They can cause a watch to function intermittently or to stop completely. If a watch with a loose-fitting hand is subjected to slight shock, the hand can swing into a new position and register incorrectly.

A careful functional examination of the hands before dismantling will frequently reveal the cause of the trouble. Once the hands have been removed the evidence is destroyed. This inspection is dealt with fully in Chapter 5 and so we will concern ourselves here with the correct fitting of replacement hands.

Seconds hand. To fit a new seconds hand to its pivot the pipe must be broached. This is done by holding the pipe in a pair of hand-tongs (Fig. 28), and enlarging the hole with a pivot broach. Only the lightest pressure is needed whilst rotating the broach between finger and thumb.

To reduce the length of a pipe, drill a hole in a strip of brass just large enough to allow the pipe to pass through. Place the hand on the filing block, pipe uppermost, and lower the brass strip over it. The exposed end of the pipe can now be filed.

The internal burr is removed by using the broach as before.

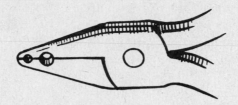

Fig. 28. Hand-tongs.

The external burr can be taken off by leaving the hand on the broach and using a pivot file as for pin-filing. (Chapter 1.)

The outside diameter of the pipe is reduced by using the same pin-filing technique.

Hour hand. The socket of the hour hand should be a push fit over the pipe of the hour wheel. To enlarge the socket hole do not use a broach because this method produces burrs. The socket is held in the hand-tongs and the hole enlarged by a rat-tail file. Alternatively, the diameter of the hour wheel pipe can be reduced in the turns. If when this is done the hour wheel pipe is found to be gripping the cannon pinion, the hour wheel pipe must be lightly broached with the hour hand in position.

It may be found that the outside diameter of the hour hand socket is too great to clear the hole in the dial. In this event the socket should be reduced in the turns.

Minute hand. Cannon pinions are either square-ended or round-ended, and the hole in the minute hand centre boss must be the same. To fit a minute hand to a square ended cannon pinion a square file is used to enlarge the hole of the centre boss. The file is used in the corners of the square hole and not on the sides, making one stroke in each corner at a time. This will assist in maintaining the square shape.

To enlarge a round hole, use a rat-tail file rather than a broach.

If the thickness of the centre boss needs to be reduced this can be done by careful filing, but a more satisfactory way is in the turns.

When opening out the holes of the hour hand and minute hand, care must be taken to keep the hole square with the hand, and not to remove too much metal. The finished result should be a push fit. In the case of the seconds hand the very lightest nip of the pipe on the pivot is sufficient.

Enamel dials. Due to the fragile nature of the layer of enamel, these dials must be treated with care. If the enamel is cracked, it is best left alone. Small pieces can be put back in position by using glue. If a small piece is missing the space can be filled in with dial enamel. The cement is warmed and run on to the dial base and allowed to set. The cement is then filed down to the level of the surrounding enamel. Hold the dial by one of the feet in a pair of pliers and pass the dial through a spirit flame. This will momentarily melt the cement which will set with a glossy surface.

To enlarge a dial hole a cone-shaped emery stick is used first to produce a chamfer. These emery sticks are made specially for this purpose and can be obtained from your supplier. The hole is then made larger by filing with a rat-tail file moistened with turpentine.

A broken foot can be replaced. The enamel at the back of the dial is removed from the affected area by scraping with a sharp pointed tool. A new foot is made from a length of copper wire and is silver-soldered to a small disc of copper a little larger in diameter than the wire. The new foot is now soft-soldered into position. Marks or discolouration are best removed with warm soapy water.

Gold and silver dials. These dials are very delicate and should

Components

100	Plate
105	Barrel bridge
110	Train wheel bridge
121	Balance cock
125	Pallet cock
180	Barrel complete with arbor
182	Barrel and cover
185	Barrel
190	Barrel cover
195	Barrel arbor
201	Centre wheel complete
210	Third wheel complete
220	Fourth wheel complete
240	Cannon pinion
250	Hour wheel
260	Minute wheel
301	Regulator index
311	Top balance endpiece
330	Bottom balance endpiece
401	Winding stem
407	Castle wheel
410	Winding pinion (crown wheel)
415	Barrel ratchet
420	Top crown wheel (transmission wheel)
422	Crown wheel collar
425	Click
430	Click spring
435	Return bar
440	Return bar spring
443	Setting bolt (pull-out piece)
445	Setting bolt spring (pull-out piece spring)
450	Intermediate wheel
705	Escape wheel complete
710	Pallets complete
721	Balance complete
723	Balance staff
730	Roller
734	Hairspring complete
738	Hairspring stud
770	Mainspring

Screws

5105	Barrel bridge (cheesehead, polished)
5110	Train wheel bridge (cheesehead, polished)
5121	Balance cock (cheesehead, polished)
5125	Pallet cock, (cheesehead, polished)
5311	Top balance endpiece (countersunk, polished ends)
5330	Bottom balance endpiece (countersunk, polished ends)
5415	Barrel ratchet
5420	Top crown wheel (transmission wheel)(left hand)
5425	Click
5443	Setting bolt (pull-out piece)
5445	Setting bolt spring (pull-out piece spring)
5721	Balance
5738	Hairspring stud
5750	Dial edge

Jewels

630	Top balance
631	Bottom balance
615	Escape top
616	Escape bottom
620	Top pallet
621	Bottom pallet
646	Entry pallet
647	Exit pallet
610	Top fourth
611	Bottom fourth
605	Top third
606	Bottom third
601	Top centre
602	Bottom centre
648	Roller (impulse pin)

KEY TO PLATE 1

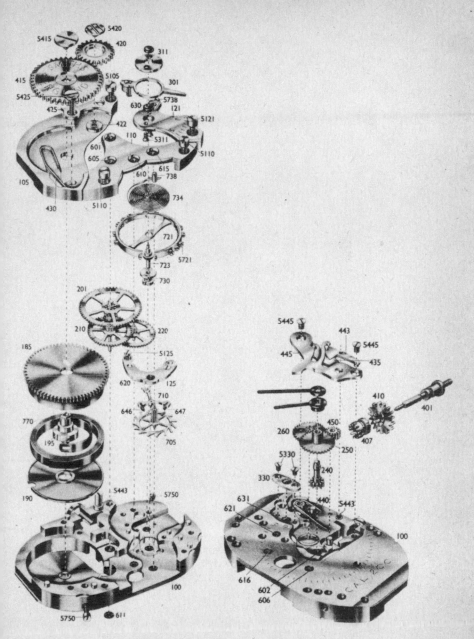

Plate I: 5¼ x 8¾ ligne 15-jewel lever movement (see facing page).

Plate II: 5¼ x 8¾ ligne movement.

Plate III: Train wheel bridge removed.

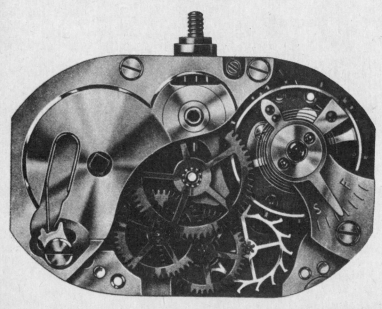

be handled in soft tissue paper. Finger marks can be removed from gold dials by light rubbing with tissue paper. Silver dials discolour badly with age. Both types of dials may be washed in warm soapy water and dried in boxwood dust. They should never be brushed or rubbed with a cloth.

Broken feet can be replaced by soft-soldering a length of copper wire into position.

Filing is the only method of enlarging holes. A rat-tail file is used inserted from the face side.

CHAPTER EIGHT

Keyless work

KEYLESS work as the name implies is a method of winding the mainspring without the use of a key.

There are a few ways in which this is done but we will consider the method most used in wristwatches.

Figure 29 shows a typical arrangement of keyless work. The winding stem is threaded at one end for the winding button. At the base of the threaded portion is a shoulder that works in the movement plate. Beneath this shoulder a groove is cut to receive the pin of the pull-out piece, and beneath the groove is another shoulder that locates the crown wheel. The remaining length is squared to take the castle wheel with the exception of the end which is turned down to form a pivot.

The pull-out piece is retained in position by a shouldered screw. The crown wheel rotates on the lower shoulder of the winding stem. The castle wheel has a square hole and although free to slide up and down the squared portion of the winding stem must nevertheless rotate with it. The check-spring is held in position at one end by a screw, and the other end is positioned in the groove of the castle wheel. The upper face of the castle wheel and the lower face of the crown wheel are cut with identically shaped ratchet teeth and are in mesh with each other. The upper edge of the crown wheel is cut with gear teeth which mesh with the transmission wheel. The bottom face of the castle wheel is cut with gear teeth and can be meshed with the intermediate wheel.

The normal position for this mechanism is as shown in figure 29. Turning the winding button and stem will cause the castle wheel to rotate. Because the check-spring is holding the castle wheel in mesh with the crown wheel the crown wheel must rotate as well.

In turning the crown wheel, the transmission wheel on the top plate is rotated which, through the ratchet wheel, causes the mainspring to be wound.

If a sharp pull is given to the winding stem, the pull-out piece will turn with its screw causing the check-spring to be pressed downward. This position is maintained by the end of the pull-out piece dropping into a step cut in the edge of the check-spring.

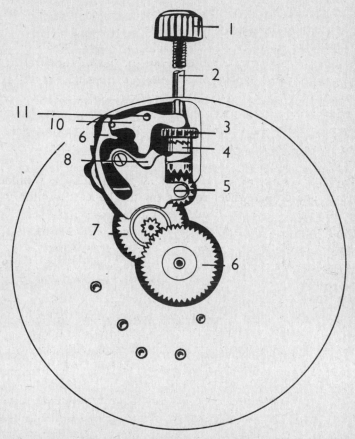

Fig. 29. Keyless work. 1. Winding button. 2. Winding stem. 3. Crown wheel. 4. Castle wheel. 5. Intermediate wheel. 6. Hour wheel 7 Minute wheel. 8. Return bar. 9. Check spring. 10. Pull-out piece. 11. Pull-out piece screw.

Pressing the check-spring downward causes the castle wheel to move down the winding stem, become disengaged from the crown wheel, and mesh with the intermediate wheel which is in mesh with the motion work.

In this position turning the winding stem will operate the motion work and alter the hand setting.

A downward pressure on the winding button will overcome the resistance offered by the check-spring and the pull-out piece and the keyless work will be restored to its original position.

If the winding mechanism is stiff to operate, inspect the ratchet wheel for freedom of movement. If the wheel is binding on the plate either the shoulder of the barrel arbor is not projecting clear of the plate or the arbor holes in the plate are worn.

In the first case, when screwing the ratchet wheel to the arbor the wheel will be pulled down on to the plate. Inspect the plate and the mainspring barrel for burrs around the arbor hole and if necessary skim the surfaces in the turns. The alternative is to fit a new arbor. If the arbor holes have become enlarged the arbor will lean over causing the ratchet wheel to rub against the plate. The remedy is to bush the arbor holes in the plate.

Slipping is caused by badly fitting wheels and badly meshed wheel teeth. Turn the winding stem in the direction of winding and observe the effect on the wheels. If the crown wheel slips on the castle wheel, the cause will be worn teeth or worn winding stem holes in the wheels or a worn winding stem. The wheels should not be loose on the stem. Renewal of parts will effect a cure.

If the up-and-down shake of the transmission wheel is excessive the action of rotating the crown wheel will cause the transmission wheel to be pushed away, until the shake is taken up, resulting in insufficient depth of mesh and subsequent slip. The transmission wheel shake must be reduced. First ensure that the transmission wheel boss is screwed down tightly. If this is in order, then the height of the wheel bearing must be reduced. Some transmission wheels are made complete with a wheel bearing in the form of a boss. The length of this boss can be reduced by stoning. If the bearing forms part of the plate, then the height of the bearing must be machined down in the turns.

Sometimes when pulling out the winding stem to alter the setting of the hands, the stem comes away from the movement. This is caused by the pin of the pull-out piece failing to retain the stem in position. If the pin of the pull-out piece, or the groove in

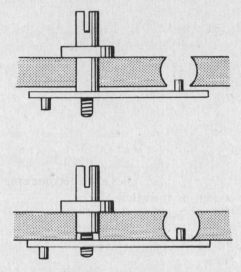

Fig. 30. Pull-out piece.

the winding stem are worn, replacement parts will be required. If the winding stem becomes loose in the plate caused by the plate being worn, the excessive side-shake of the stem will result in partial disengagement of the pin of the pull-out piece in the winding stem groove. In such cases the plate must be bushed or an oversized winding stem fitted.

Another cause of the winding stem coming away is the pull-out piece becoming loose on the pull-out piece screw. When correctly fitted the screw should be tight in the pull-out piece but free to move in the plate so that when the winding stem is pulled out or pushed in, the pull-out piece and its screw rotate as one.

When the pull-out piece screw is in position make sure that the shoulder nearest the thread projects beyond the face of the plate (Fig. 30, *a*). If it does not, it means that when the pull-out piece screw is threaded into the pull-out piece, the pull-out piece will be held against the plate before the screw is fully tight (Fig. 30, *b*).

CHAPTER NINE

Barrels, mainsprings and fusee chains

BECAUSE the mainspring and barrel assembly operates very slowly, it must be free-moving and smooth-acting. There must be no possibility of rubbing or tendency to stick.

If rubbing has taken place, an inspection of the barrel and the watch frame quite often reveals the tell-tale marks.

Let us first make sure that the barrel rotates freely and squarely about the arbor.

Barrels. Place the arbor in the barrel and snap the cover on. The arbor should have slight end-shake.

Now hold the end of the arbor in the jaws of a pin-vice with the cover uppermost (Fig. 31). The arbor should be held at an angle so that the gear teeth on one side are just clear of the jaws

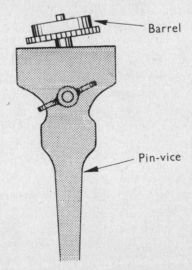

Fig. 31. Checking barrel for running true.

of the pin-vice. By rotating the barrel it will be seen whether the barrel is running true.

If the barrel wobbles, then the arbor hole in the barrel or the cover needs bushing. Enlarging the hole to take the bush is best done on a lathe, and fitting the bush is carried out with the aid of a watchmaker's punch and stake. If you do not possess this equipment then the work will have to be sent away.

With the barrel free and square about the arbor, fit the barrel between the plates. There should be no end movement of the arbor between the plates. All that is required is freedom of the arbor to turn when the spring is being wound.

A loose barrel cover can be tightened by placing it on a flat steel stake and gently tapping all round the edge with the flat of a hammer. The hammer face must meet squarely with the cover if marks and bruises are to be avoided. Use a stroking action by bringing the hammer downward and outward. After each tap, rotate the cover a little until the place of starting has been reached. This will spread the metal until a good fit is obtained.

To fit a new mainspring hook to a barrel, first drill a hole in the side of the barrel at a slight angle pointing away from the direction of pull of the mainspring. The hole is then threaded slightly undersize.

Taper a piece of steel wire by filing and cut a thread with a screw plate, holding the wire in a pin-vice. Remove the pin-vice, cut the wire close to the screw plate and shape the protruding piece of wire into a hook by filing.

Unscrew the threaded wire from the screw plate and insert the small diameter end of the wire into the hole in the barrel entering from the inside.

The protruding wire on the outside is now held by the pin-vice and the wire is screwed into position by unscrewing from the outside.

The hook should project into the barrel a distance no greater than the thickness of a piece of the mainspring.

When in place remove the pin-vice, cut off the wire on the outside, file down close to the barrel and finish with an oilstone slip.

Similarly, new hooks can be fitted to barrel arbors. A hole is drilled in the arbor and a steel pin is driven in tight. The new pin is shaped by filing and finished with an oilstone slip.

Occasionally a barrel tooth becomes bent as the result of a mainspring breaking. The bent tooth can often be straightened by inserting the blade of a knife at the side of the bent tooth and gently levering back into position.

Mainsprings. If a mainspring breaks close to the outer end, it can be repaired. A break anywhere else necessitates the fitting of a new spring.

When a replacement mainspring is needed, it is better to purchase a spring from the manufacturer of the watch. The length of the spring will be correct and it will be hooked and ready for fitting into the barrel.

The alternative is to purchase a suitable spring, cut it to length and shape the outer end. Assuming the spring to be replaced is the correct gauge, width and length, then the purchase of a replacement should present no problems.

However, it is as well to know a little about the design of a mainspring and a barrel assembly in case the spring to be replaced does not conform to the specification laid down by the watch manufacturer.

Two identical springs, but of different lengths will, if bent to the same curve, apply different forces. If one spring is half the length of the other it will apply twice the force.

A long spring will take up more space in the barrel and it therefore cannot be wound as many turns as a shorter spring. This will result in the watch unwinding itself too quickly.

A spring that is too short will when wound exert too great a force and cause a high rate of wear in the wheel train and escapement.

A simple way to establish how long the spring should be is illustrated in figure 32. Place the barrel on a sheet of metal and with a sharp pointed tool scribe a circle using the barrel hole as a guide. Mark the centre of the circle. Take a pair of dividers, place the point of one leg on the centre of the circle and adjust the dividers until the other leg just touches the base of the barrel wall (Fig. 32, *left*). Measure this distance and readjust the dividers to two-thirds of the original distance. With the barrel still in position on the sheet of metal lightly scribe a circle on the barrel inner face (Fig. 32, *centre*).

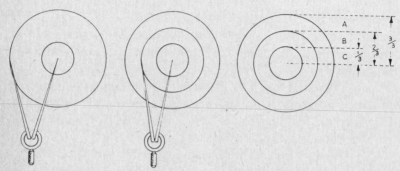

Fig. 32. Marking barrel to show correct mainspring length. A. Space for unwound spring. B. Space for wound spring. C. ½ diameter or arbor.

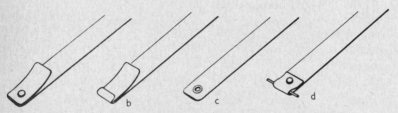

Fig. 33. Methods of hooking outer ends of mainsprings.

When the new spring is in the barrel, it should take up the space between the barrel wall and the scribed line.

The force of a spring also varies with the thickness and the height. If it is twice the thickness the force is increased eight times. A spring which is twice the height will have twice the force.

If the spring is too thick it will not be possible to obtain the required number of turns. If the height is too great the spring will foul the barrel and the cover, and if the height is insufficient then the spring will buckle and once again foul the barrel and the cover.

If the inner face of the barrel cover is recessed, the height of the mainspring should be level with the shoulder in the barrel wall upon which the cover rests. If the cover is not recessed then the height of the spring should be just below the shoulder.

There are various methods of hooking the outer ends of main-springs to barrels and figure 33 illustrates those in common use.

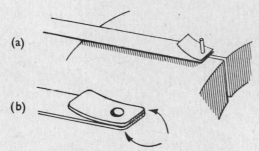

Fig. 34. Fitting a riveted hook.

To fit a riveted hook, soften the extreme end of the spring in a spirit flame by heating to a black heat. Hammer the end of the spring flat and drill a small hole to take a rivet. Select a short piece of spring the same width and thickness as the mainspring and drill a hole in one end using the same drill as before. File the other end square with the sides. Hold the two pieces of spring together and broach out both holes as one. Remove the burrs and polish both pieces with fine grade emery paper working the grain down the length of the spring.

Cut off a short length of steel wire, hold it in the pin-vice and file it to fit the hole. Push the wire into the hole and grip the wire in a vice (Fig. 34, *a*).

Cut the wire close to the face of the spring and rivet over with a round-face hammer.

Remove the pin from the vice and cut off the other end as before. Place the riveted head in a riveting stake or dolly and complete the operation by riveting the other end. File the end as shown in figure 34, *b*, and remove the burrs. Polish with a fine grade emery paper.

A very narrow spring such as would be found in a flat watch would be weakened by drilling a hole. An alternative method of hooking such a spring is shown in figure 35.

Break off the end of the spring to within two or three inches of the required length. Heat the spring, at about the point where the end is to be, in a spirit flame and bend upward and over using

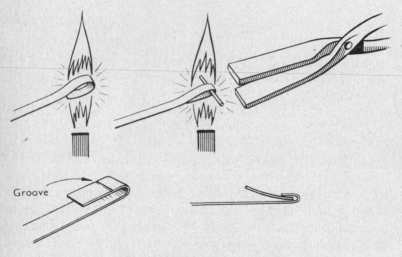

Fig. 35. Making a loose-piece.

flat-nose pliers. The pliers should be pre-heated to prevent the heat of the spring being absorbed by the cool pliers. Keep the spring in the flame whilst bending is taking place and do not force the spring. Once the correct temperature has been reached, bending will be found easy. Take the piece of broken spring and place a short length in the loop of the mainspring. Reheat and slowly close the mainspring loop over the short length of spring and gently but firmly squeeze in to shape. Withdraw the short length of spring.

Cut a groove across the face of the short arm of the loop with a file, bend back and snap off. File the end square and remove the burrs.

From the short length of spring cut a piece long enough to form the hook, file the ends square and remove the burrs.

Polish both springs with fine grade emery paper and push the short length of spring into the hooked end of the mainspring.

To make an eye, soften the extreme end of the spring by heating in the spirit flame. Drill a hole in the end and then enlarge it with a broach to fit loosely over the barrel hook. The last few cuts of the broach should be made at an angle to correspond with the angle of the barrel hook. The burrs produced by the broach must now be removed.

The end of the spring has to be curved to the same radius as that of the barrel. This will ensure that the spring is correctly hooked.

Place the spring, outer face downward, on a block of lead, and tap the inner face with a watchmaker's hammer. This will decrease the radius of the curve.

Finish off by polishing the end of the spring with fine grade emery paper.

The brace method of hooking a mainspring is used in almost all American watches. The brace consists of a short length of spring, the same width as the mainspring but slightly thicker. It carries two pivots that project into a hole in the barrel base and another hole in the barrel cover.

The brace is secured to the end of the mainspring by a rivet. It is better to purchase a new brace rather than make one. Replacement braces are supplied drilled and complete with a rivet. When the spring is fully wound, the brace swinging on its pivots, moves across the barrel.

Make sure that the pivots do not project beyond the barrel or the barrel cover. The width of the brace should not be greater than the width of the mainspring because fouling may occur between the edges of the brace and the faces of the barrel and the barrel cover during unwinding of the spring.

Fusee chains. Hold the broken end of the chain and insert a thin blade beneath the link plate. Gently lever the plate away from the rivet until the outline of the rivet can be seen.

Place the chain on a graduated stake and with a blunted sewing needle held in a pin-vice push the rivet out. It may be necessary to tap the pin-vice with a hammer.

Treat the other broken end in the same way.

Select a sewing needle from which to make the new rivet and heat it to a blue. File the needle almost to size with a very fine file and finish off with an oilstone slip.

Fit the ends of the chain into each other, making sure that the two end hooks are on the same side, and insert the needle into the rivet holes. Replace the chain over the stake and gently tap the needle home.

Cut both ends of the needle off close to the side plates and file flush. Lay the chain on its side on a metal stake and gently rivet each end of the needle with a round-face hammer.

In carrying out this repair, frequent checks are necessary to ensure that the centre plate on one end of the chain is free to swing between the side plates of the other end of the chain.

CHAPTER TEN

Escapements

THE two most popular forms of escapement today are the lever and the pin-pallet.

Whereas the lever has great precision and can be relied upon to function accurately over long periods, the pin-pallet is a cheap and robust product that lends itself well to mass production. Even so, the good service obtained from a pin-pallet escarpment cannot be ignored.

The lever (Fig. 36). To enable faults in a lever escapement to be identified a working knowledge of its function is necessary.

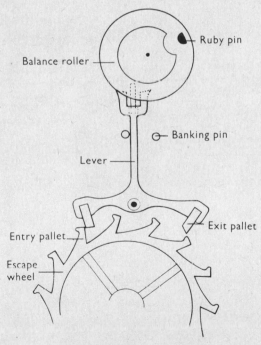

Fig. 36. Lever escapement.

We have seen how power from the mainspring, passing through the train of wheels reaches the escape wheel. We have learned that it is the pallets of the lever, interrupting the teeth of the escape wheel, that prevent the wheel train from rotating at a high speed during the unwinding of the mainspring.

The escapement consists of the escape wheel, the lever, two banking pins and the balance. Figure 37 illustrates the lever in detail and figure 38 shows the roller which is fitted to the balance staff.

In figure 39 we can see the stages through which the escapement passes in order to release the escape wheel one tooth at a time.

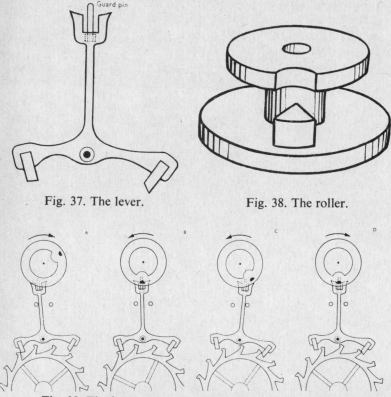

Fig. 37. The lever.   Fig. 38. The roller.

Fig. 39. The lever escapement—sequence of movements.

The escape wheel tooth hits the locking face of the entry pallet stone and the escape wheel is brought to rest (Fig. 39, *a*). The power in the escape wheel causes the tooth to press against the locking face of the entry pallet stone pulling the stone down. This downward movement is called the draw (Fig. 40). The draw pulls the lever hard against the banking pin and holds it there. This last movement of the lever is known as the run to banking. If the watch received a shock sufficient to overcome the draw, the lever

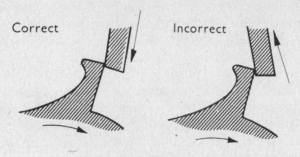

Fig. 40. The draw.

would no longer be held against the banking pin and would be free to pass across to the other banking pin. This would cause the watch to stop. To prevent this happening, a guard pin is fitted at the end of the lever which, under conditions described above, would come in contact with the roller and further movement of the lever would be checked.

To allow the lever to pass under normal conditions, a groove is cut in the edge of the roller level with the ruby pin.

The balance completes its swing, stops and reverses under the tension of the balance spring. The ruby pin enters the notch of the lever and moves the lever away from the banking pin. This causes the entry pallet stone to be pulled away from the escape wheel thereby releasing the tooth (Fig. 39, *h*).

The ruby pin continues to move round pushing the lever farther over and the exit pallet stone moves in towards the escape wheel. The adjacent tooth strikes the exit pallet stone on its locking face and the lever is held firmly against the other banking pin (Fig. 39, *c*).

The balance completes its swing and reverses direction. As the ruby pin re-enters the notch in the lever, the lever is moved away from the banking pin, pulling the exit pallet stone away from the escape wheel tooth. The escape wheel tooth, being freed from the locking face, now presses against the impulse face of the exit pallet stone. It is this impulse, transmitted through the lever, that gives a further swing to the balance (Fig. 39, *d*). While the exit pallet stone is being pushed away, the entry pallet stone is moving in to stop the next tooth and the cycle of operations repeats itself. For such small surfaces to perform these operations efficiently over long periods calls for a high standard of workmanship. Slight damage or wear can destroy the timing of the escapement or cause the movement to stop altogether. Let us then examine the escapement to detect any faults that may be present. Check the balance staff end-shake by holding the balance arm in a pair of tweezers and moving the balance wheel up and down. To know when the end-shake is excessive calls for experience, but if you remember that providing the end-shake is just sufficient to retain a film of oil, then any excess is undesirable.

Another test for end-shake is to wind the mainspring a few turns and set the movement in motion. Apply light pressure to the balance staff end-stone with a piece of pegwood and if this stops the balance, the end-shake is insufficient. Hold the watch movement between the thumb and forefinger of one hand and place the tip of the forefinger of the other hand lightly on the rim of the balance wheel. Very little movement of the finger is required to rock the balance.

Examine with the aid of an eye-glass the side-shake of the top and bottom pivots.

It is almost certain that wear will be the cause of excessive shake and the remedy is to fit a new staff and to renew the pivot bearings.

Now remove the balance cock and lift out the balance. In this condition the pallet staff can be checked for shake, the same remarks applying here as for the balance.

We now turn our attention to the locking of the escape wheel teeth by the pallet stones. First, let the mainspring down. Now fold a small piece of tissue paper and place it under the guard pin of the lever; this will hold the lever in any position we like to put it.

Take a pointed pegwood stick and position the escape wheel and lever so as to bring a tooth in contact with the impulse face of the entry pallet. Very slowly move the escape wheel forward until the tooth is just clear of the pallet. In this position the exit pallet should be locking a tooth against its locking face.

Now raise the exit pallet stone slightly and move the escape wheel forward so that the tooth comes on to the impulse face. Make sure that in this position the entry pallet stone moves in to lock the next tooth. This test must be repeated for all fifteen teeth. If the entry stone mislocks on any tooth the cause must be investigated.

Examine the tips of the escape wheel teeth. If they are showing obvious signs of wear, the escape wheel must be renewed.

Examine the escape wheel pivot bearings and the pallet pivot bearings for wear. Any of these faults will cause the escape wheel and the pallets to function at too great a distance from each other.

If no fault is found here, then adjustment must be made by altering the positions of the pallet stones.

The stones are held in the pallets by shellac. If the shellac is warmed, the stones can be moved and held in their new positions until the shellac has set. If the entry stone is pulled out, the exit stone must be moved in by the same amount, otherwise the exit stone will be too deep when locking.

Remove the lever from the watch and examine the impulse faces of the pallet stones. If they are pitted they must be replaced by new stones.

Place the lever on a blueing pan together with a small chip of shellac. Heat the pan in a spirit flame until the chip of shellac is softened. This will be the correct temperature at which to move the stones.

Leave the lever on the pan and, holding the lever in a pair of tweezers, adjust the stone with a second pair of tweezers (Fig. 41).

In many watches the undersurface of the pallets carries a film of shellac. If this film is disturbed when adjustments are made, be sure to make it good before replacing the lever in the movement.

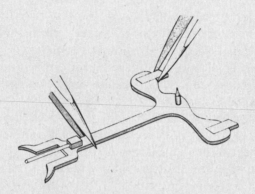

Fig. 41. Adjusting pallet stone.

If new pallet stones have to be fitted, the lever must be sent to the supplier with the order.

Place the lever on the blueing pan and hold it over the spirit flame. Insert the new stones the right way up. Put a small chip of shellac on the back of each stone and let it run over the face. Move the stone in and out of the pallet allowing the shellac to run in the join. If most of the shellac runs into the join, add another small piece to run over the face to restore the film. Fit the lever into the movement and test the locking. When the locking is correct, lift out the lever and remove any surplus shellac but leave the film intact.

The next test is to see whether the lever has the correct run to banking. It is this free movement of the lever that makes draw possible.

A few turns are given to wind the mainspring. The lever is wedged with tissue paper as before and very slowly moved to one side until a tooth of the escape wheel just touches the locking face of one of the pallet stones. At this point the distance between the lever and the banking pin should be about the thickness of a thin razor blade. Both sides of the lever should have the same distance.

If the run to bank is too great, the distance between the banking must be reduced. If banking pins are employed, the pins are bent towards each other and then given a second bend to bring them parallel (Fig. 42). This is important, otherwise if the pins were left tapering towards each other the run to banking would be further reduced when the movement was turned over.

Some American watches use pins placed eccentrically in screws. All that is needed here is that the screws are turned until the pins are in the correct position in relation to the lever. The problem of maintaining the pins in a vertical attitude does not arise.

Solid bankings are employed in some movements and with these a little extra work is required to make adjustments.

To decrease the gap a slot must be cut in each banking with a very thin file such as one used for deepening the screwdriver slots in screw heads (Fig. 43, *a*). The two pillars thus formed are bent inward and their inner faces filed vertically.

Increasing the gap of solid bankings necessitates the removal of a thin shaving of metal from the inner face using the blade of a sharp screwdriver (Fig. 43, *b*).

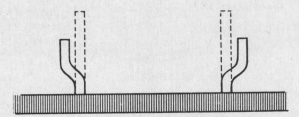

Fig. 42. Adjusting banking pins.

Now let us look at the draw to see that we have the right amount. Remove the tissue paper wedge from the lever, and slowly move the lever to one side until the locking face of one of the pallets is just clear of the escape wheel tooth.

At this point release the lever which should jump across to the banking. Both pallets must be tested in the same way.

If the lever is sluggish when moving indicating insufficient draw, the attitudes of the stones in the pallets must be altered.

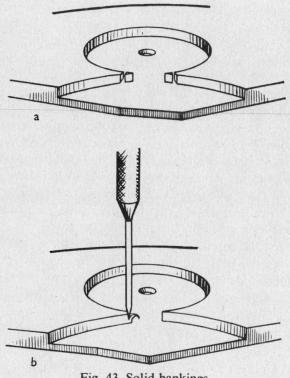

Fig. 43. Solid bankings.

Let the mainspring down and lift the lever from the movement. Place the lever on the blueing pan with the shellac side uppermost together with a chipping of shellac. Hold the pan in a spirit flame and heat until the piece of shellac becomes soft. Place the pan on the bench and hold the lever with tweezers. Push the pallet stone to one side in the direction required and hold it there until cooled.

Replace the lever in the movement, wind the mainspring a few turns and test again. If the draw is still insufficient another pallet stone, slightly narrower must be fitted.

Sometimes a lever movement stops each time it is shaken or receives a slight shock. If examination reveals the lever at rest on the wrong side of the roller, the fault is a short guard pin. Shaking the movement causes the lever to pass the roller in the wrong direction and when the balance reverses its swing the ruby pin comes to rest on the edge of the lever horn instead of entering the notch.

The remedy is to lengthen the pin and this can be done either by bending it forward or by fitting a new one.

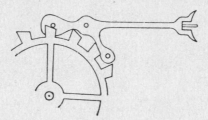

Fig. 44. Pin pallet

The original pin is pushed out in the direction of the pallets. A new brass pin with a long taper is made by the pin-filing method. Smooth the pin with the finest grade emery stick and finish with the flat burnisher.

Insert the new pin from the pallet end of the lever. Cut off the surplus length at the back with the nippers, and file the operating end of the pin to the required length. Finish off the end with an Arkansas stone.

Pin pallet (Fig. 44). In this form of escapement the pallets are fitted with pins instead of stones, and the impulse faces are cut on the ends of the escape wheel teeth. The draw of the escape wheel teeth on the pallet pins causes the pins to bank against the bottom of the teeth, it is therefore quite usual for some manufacturers not to fit banking pins with this type of escapement.

Mislocking is a common cause of failure in this type of escapement but it is quite easy to test and the remedy is simple.

Give a few turns to the mainspring. With a pointed pegwood stick rotate the balance very slowly until an escape wheel tooth is released. In this position make sure that the other pallet pin has locked another tooth. For the locking to be correct the locking face of the tooth must be above the centre of the pallet pin (Fig. 45, *a*). Incorrect locking is illustrated in figure 45, *b* and *c*.

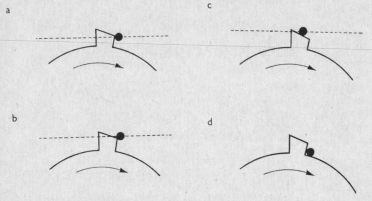

Fig. 45. Pin pallet locking. (a) Correct. The escape wheel tooth strikes the pallet pin above the centre. (b) Incorrect. The escape wheel tooth strikes the pallet pin below the centre. (c) Mislock. The pallet pin is struck by the impulse face of the escape wheel tooth. (d) Draw. The pallet pin is pulled down to the base of the tooth.

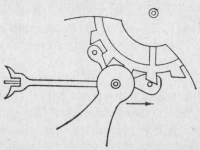

Fig. 46. Direction of bending the tongue to overcome mislocking.

Now carefully rotate the balance a little farther in the same direction and watch the draw. The tooth should pull the pallet pin down to the base of the tooth (Fig. 45, *d*).

Repeat this test on both pallet pins and for all fifteen teeth of the escape wheel.

If mislocking occurs anywhere the pallets must be brought closer to the escape wheel. This is done by slightly bending the

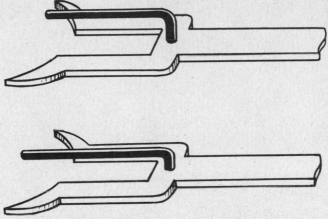

Fig. 47. Lengthening the guard pin.

tongue that carries the pallet pivot bearing hole. The tongue is held by the long nose pliers and bent sideways towards the escape wheel (Fig. 46).

After bending repeat the above tests.

To test the draw of the escape wheel teeth rotate the balance until the impulse pin is clear of the lever. Hold the balance in this position and move the lever against the roller. If the escapement has sufficient draw, the lever will jump back into its original position when released. This test must be carried out on both pallet pins and with all fifteen teeth of the escape wheel. If the draw is weak the remedy is a replacement wheel.

The amount of shake on the roller needs to be checked. By moving the balance round so that the impulse pin on the roller is clear of the notch in the lever, the shake can be tried. Now swing the balance in the opposite direction and try the shake again. Both should be equal.

If it is found that there is more shake on one side, the lever must

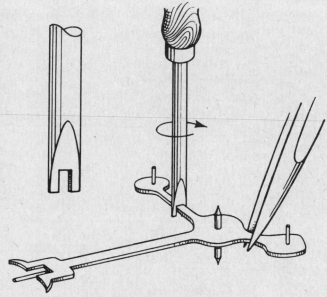

Fig. 48. Closing the pallet arms.

be removed and bent sideways in the direction of reducing the greater shake. The lever is placed on its side on a stake held in the vice. A chisel with the cutting edge rounded off is placed on the upper side of the lever and a light tap given with a hammer. This will cause the lever to bend upward.

Sometimes it is found that both sides have no shake at all, in this case the length of the guard pin must be reduced.

If the shake is excessive on both sides the guard pin must be increased in length. This can usually be done by first straightening the pin and then re-bending it as close to the lever as possible (Fig. 47). In some watches the guard pin is cut from the solid in which case the metal has to be stretched by light tapping with a hammer, with the guard pin resting on a metal strip held in the vice.

After any adjustment replace the lever, re-check the shake, and make sure the guard pin engages on the roller and that the impulse pin engages freely in the lever notch.

The distance an escape wheel tooth moves before being stopped by a pallet pin is known as the drop. This distance must be the same on both pins.

Move the balance round slowly as before and note the amount of drop on the pin. Reverse the direction of the balance and note the amount of drop on the other pin.

If one side is greater than the other, then the pallet arm on the greater side must be bent toward the other arm.

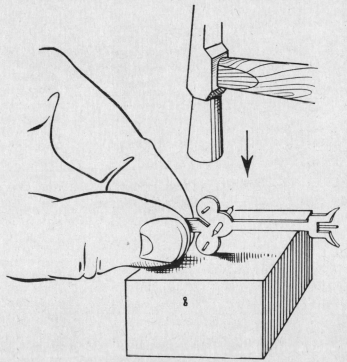

Fig. 49. Closing the pallet arms.

Remove the lever and bend the arm sideways using the brass faced pliers or a made-up tool as shown in figure 48. If the pallets have no arms, a slot must be cut in the pallet and then carefully closed by tapping with a hammer (Fig. 49).

If the pallet pins are badly worn they must be renewed before making adjustments to the escapement.

Place the lever over a graduated stake and with a length of blunted sewing needle gently hammer the old pins out.

The replacement pins can be made from a sewing needle. Select one that is slightly oversize, cut two pieces to the length required and carefully hammer them into the pallets over the stake (Fig. 50).

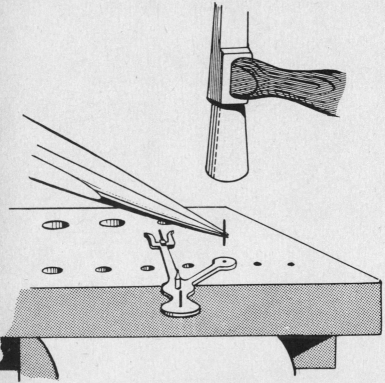

Fig. 50. Fitting replacement pin.

CHAPTER ELEVEN

Balances

A WATCH balance consists of a hairspring and collet assembly, a balance wheel and a roller, all of which are fitted to a common staff.

If a balance is subjected to a rise in temperature, the wheel diameter will be increased causing the bulk of the weight to move away from the centre. This will have the effect of slowing down the movement. A drop in temperature will have the reverse effect and the movement will gain.

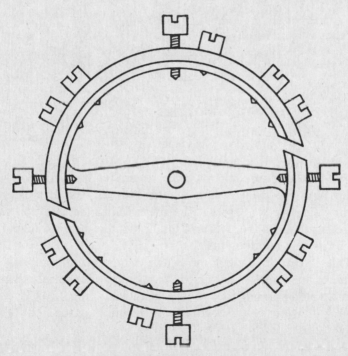

Fig. 51. Compensated balance.

To overcome this, the more expensive watches are fitted with a means of compensating for variations in temperature.

Compensated balance (Fig. 51). The wheel is made of steel but the outer edge of the rim is faced with brass. This is known as bi-metallic.

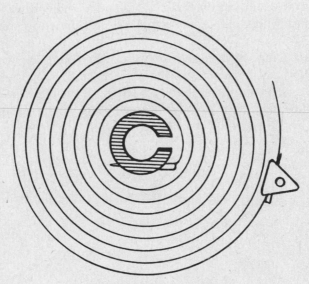

Fig. 52. The hairspring.

The rate of expansion of brass is greater than that of steel when both are subjected to identical conditions. If a straight bi-metallic strip was heated, the brass would increase its length more than would the steel and the bi-metallic strip would curve, the brass being on the outside.

This is the principle upon which temperature compensated balances are made. With a rise in temperature the wheel expands, which would normally slow down the rate of swing. In expanding, the two halves of the rim curve inward bringing the bulk of the weight closer to the centre.

If the effect of inward curvature balances out the effect of expansion, the watch will maintain a constant speed.

Provision is made for fine adjustment by fitting screws into the wheel rim.

Usually it is not necessary to make any adjustment here after the screws have been set by the manufacturer.

In addition to the compensating screws, there are four quarter screws placed equidistant from each other. These screws have longer threads than the compensating screws and are used to change the speed of the balance.

Unscrewing an opposite pair will slow down the movement and screwing them in will increase the speed. This is because the distance from the centre to the bulk of the weight is being altered.

Hairspring (Fig. 52). The hairspring is made from steel wire rolled flat and thin and coiled into a spiral. The strength of the spring will vary according to the width and thickness of the material. The inner end is pinned to a collet and the outer end is pinned to a stud.

Most collets are made of brass and shaped in the form of a split collar or bush. The centre hole is made to fit tightly over the balance staff. One side of the collet is split to provide springiness. Some of the best English lever watches have steel collets, they

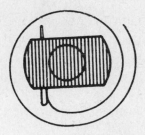

Fig. 53. Steel collet.

are shaped with two flats, one on each side, and are not split (Fig. 53). The clearance between the flats and the inner coil of the spring enables truing of the hairspring to be made more easily.

There are different types of studs in use. One of the most common types in use is a brass block with a pin machined on one face, the pin being either a push fit in the balance cock or held in position by a screw entering the balance cock at the side.

The balance staff moves in two jewel bearings, one in the bottom plate and the other in the balance cock.

A regulator or index is fitted to the balance cock and is free to move within a restricted arc. The hairspring passes between two index pins in the index.

When the index is moved from one position to another the effective length of the hairspring is altered causing the movement to gain or lose according to the direction in which the index was moved.

Fitting a flat hairspring. The original hairspring with its collet must first be removed. Hold the balance between finger and thumb at each end of the balance arm; holding it this way will prevent distortion. Push the blade of a steel oiler into the slot of the collet, this will spring the collet and reduce the grip on the balance staff. Turn and lift the oiler in one movement and the collet will be freed from the staff.

If the slot in the collet is too wide for the oiler, a thin knife blade inserted beneath the collet will lift it from the staff.

It is not advisable to hold a stock of hairsprings because of the very wide range in use in present-day watches.

Springs are made in varying strengths in different sizes and to enable your supplier to select the correct spring it will be necessary to send the count (number of vibrations in one hour), the balance and the balance cock with your order. We have seen how the count is calculated in Chapter 6.

On receipt of the spring we must ensure the strength is correct and to do this the spring is temporarily fitted to the balance and a count is made.

First, lay the spring on the underside of the balance cock centrally over the pivot hole. The diameter of the spring must be sufficient to allow the outer coil to pass through the index pins of the balance cock and then be pinned in the stud. If the spring is too large make a note of the approximate position that will be pinned to the stud.

Plate IV: Wheel train and lever in position.

Plate V: Crown wheel
and castle wheel.

Plate VI: Winding mechanism.

Plate VII: Smiths striking clock.

Lower the spring over the balance staff and hold it temporarily in position by fitting the collet above it. Grip the hairspring in tweezers at the position where the spring passes through the index pins. Lift the balance wheel and spring so that the wheel hangs under its own weight. Lower the balance until the staff just

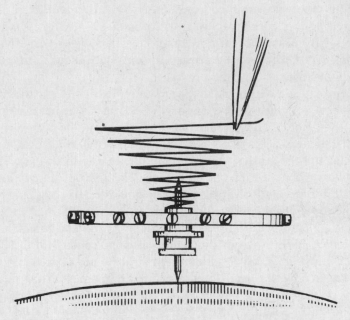

Fig. 54. Method of counting vibrations.

touches the glass of a watch that is fitted with a seconds hand. With the finger and thumb of the other hand rotate the balance wheel about half a turn and then release it. Count the number of double vibrations made in one minute (Fig. 54). If the correct spring has been supplied the count will be the same as was calculated for the gear train. Remove the collet and then the spring from the balance staff.

We are now ready to fit the spring to the collet, but first we will need a tool to enable us to break off the surplus length of spring.

Drive the point of a sewing needle into a slender wooden

handle and stone away the top of the needle eye leaving a fork end (Fig. 55).

Lay the collet on a sheet of stout white paper and place the spring centrally over the collet. Hold the spring in a pair of tweezers about a quarter of a turn from the centre. Place the bending tool over the spring and close to the tweezers. Bend the short length of spring backward and forward a few times and it will break away. Repeat this process until sufficient length of spring has been removed from the centre to allow it to pass freely over the collet.

Bend the end of the spring inward using two pairs of tweezers and then straighten it. The length of the spring to be bent inward is a little more than the length of the pin hole in the collet.

The pin that is going to secure the spring to the collet must now be made. Take a piece of brass wire, file a taper to fit the collet and burnish it with the flat burnisher. File a flat down one side of the pin a depth of approximately one-third of the diameter of the pin. Polish the flat with the finest emery stick.

Straighten one of the pieces of broken spring and insert it in the collet. Push the taper pin tightly into the collet with the flat side against the piece of spring and mark the pin at each end of the pin hole with a knife. Remove the pin and cut off the piece between the two knife marks. Polish the ends to remove all burrs.

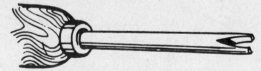

Fig. 55. Tool for breaking off end of hairspring.

Refit the pin and the piece of spring to ensure that the pin finishes a little more than half-way through the hole.

A collet holder must now be made. A piece of steel rod is filed at one end to produce a long taper. Polish the taper with a fine emery stick and finish with a burnisher.

Lower the collet on to the taper until it just grips. Make sure it is square with the holder and not at an angle.

Hold the hairspring in tweezers, lower it over the collet, insert the inner end of the spring into the collet keeping the spring level and central with the collet.

Insert the taper pin into the collet until the spring is just gripped. Push the collet from the holder and lay the spring on the white paper. In this way the coils of the spring can be checked

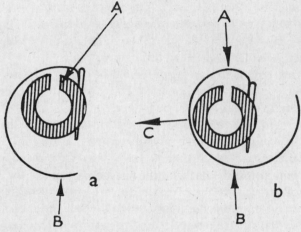

Fig. 56. Centralizing the hairspring to the collet.

to make sure that they are central about the collet. If they are not central the spring must be carefully bent to make them so (Fig. 56). All bending should take place within the first quarter coil from the collet.

In figure 56, *a*, an oiler is inserted at *A* and given a twist. This will push the coil away from the collet at that position but at the same time bring it closer to the collet at *B*. In figure 56, *b*, the diameter of the coil is reduced by applying tweezer pressure across *A* and *B* causing the coil to move away from the collet at *C*.

Before making any alterations study the contours of the spring carefully and decide exactly where the spring needs bending to produce the desired result. Two pairs of fine pointed tweezers are best for this operation.

To check for flatness of the spring the balance is mounted on an arbor and is spun between callipers or turns. If the spring is out of true it can be corrected by gripping it with tweezers close to the collet and turning in the required direction. Providing the pin was not put in too tightly the pin will move round with the spring. Slowly spin the balance after making each adjustment to check the result. When the balance can be rotated once with no variation of the spring, the arbor can be removed.

Press the pin firmly into the collet using a strong pair of tweezers. It is essential that there should be no movement of the spring in the collet when the pinning is finished.

The next step is to pin the outer end of the spring to the stud. Make a taper pin in exactly the same way as we did for the collet, but this time the taper must be longer.

Move the index to the midway position. Lay the balance cock on the bench with the index pins uppermost. Place the spring on the balance cock with the collet centrally over the pivot hole and note the position where the stud will be. Break off the surplus length of spring allowing for a little to protrude beyond the stud.

Remove the spring and fit the stud to the balance cock. Replace the spring on the balance cock and pass the end through the pin hole in the stud. Push the pin into position. Examine the assembly closely to make sure that the spring is supported parallel to the balance cock. If it is not, turn the pin which is still in the pin-vice and the spring will move with it. When the spring is parallel, make the pin a short distance from the stud with a knife and break the pin off. Push home tightly with strong tweezers.

The position and shape of the spring may now need altering. The outer coil must lie naturally between the index pins without side pressure or side play. If necessary the pins must be bent closer together. The second coil must not foul the inner index pin or stud, and the collet must rest centrally over the pivot hole in the balance cock.

It is sometimes found that the position of the stud is a little too far from the centre and this causes the spring to be out of centre. To rectify this fault bend the spring close to the stud at *A* in figure 57. If the spring touches one of the index pins, bend the spring at *B* in figure 57 until the coil lies naturally between the pins.

The balance can now be fitted to the movement, the movement set in motion and a final check carried out to ensure that the spring is functioning correctly and that no fouling is taking place.

Repairing a hairspring. Once a spring is damaged the most satisfactory repair is to fit a new one, but there are circumstances when an attempt to straighten a bent hairspring is justified.

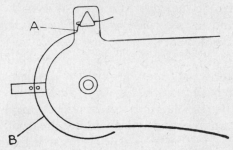

Fig. 57. Centralizing the hairspring and collet to the jewel hole.

Sometimes a spring becomes entangled. If it is caught on the stud careful manipulation will often restore the spring to its correct position. If this is not possible, the pin must be removed from the stud, the spring withdrawn and, commencing from the centre with a needle, the spring must be followed throughout its length disentangling as you go.

If the spring has jumped over the rim of the balance wheel, the spring must be released from the stud, the collet must be removed from the staff and the spring rotated in order to unwind itself from the balance.

A spring can be out of true either in the round or in the flat. That is to say the coils no longer conform to their original shape or the spring is no longer flat.

Remove the spring complete with collet and lay it on a sheet of stiff white paper. Follow the coils round with a needle starting from the centre. At the first irregularity of curvature bend the spring correctly with the point of a needle holding the spring in tweezers. Continue this process until the outer end of the spring has been reached.

Now the spring must be checked for flatness. Hold the spring up and look on its edge. Any upward or downward bends will be immediately apparent. Once again starting from the centre note each irregularity and bend the spring back into shape using two pairs of tweezers.

Having restored the flatness of the spring we must now be sure that it lies parallel to the collet. To check this, fit the collet to a turning arbor and rotate the arbor in the turns. Correction can be made by a slight twist to the spring as close to the collet as possible.

All that remains now is to check that the spring is parallel and central with the balance cock. Fit the spring to the stud, ensure that the outer coil rests naturally between the index pins and that the collet is central over the balance staff pivot bearing.

Renewing index pins. If a hairspring is subjected to shock it sometimes jumps from between the index pins or the second coil jumps in. To prevent this, the index pins should be as long as possible. If the tops can be turned inward slightly so that they both meet then isolation of the outer coil will be maintained.

Old pins are cut off and filed down flush. The stumps are driven out by a blunted needle with the index on a stake. The new pins are made with a very slight taper, polished with the fine emery stick and finished with a burnisher. The pins are then pushed tightly into the index and cut off to length.

CHAPTER TWELVE

Shock proofing

THE parts of a watch most likely to suffer from shock are the pivots of the balance staff. The diameter of these pivots is approximately the same as that of a human hair. To knock or drop a watch imparts shock to the movement and invariably some damage is caused to the balance staff pivots.

If the shock is severe, at least one pivot will break off and the watch will stop. A less severe shock is likely to cause one or both pivots to bend. This damage may not be sufficient to prevent the movement from functioning but it most certainly can cause bad

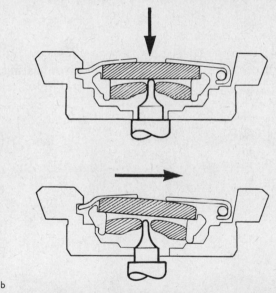

Fig. 58. Diagrammatic sketch of Kif Flector device. (a) Axial shock. (b) Radial shock. Under the effect of a shock the balance wheel is displaced, its pivot carrying away the jewel in-setting and end-stone. The shock is absorbed by the part of the staff or the arbor which hits the block. After the shock the spring instantly replaces the whole in its initial position.

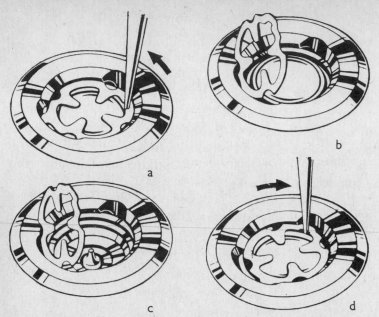

Fig. 59. Kif Flector. Removing end-stone and setting jewel.

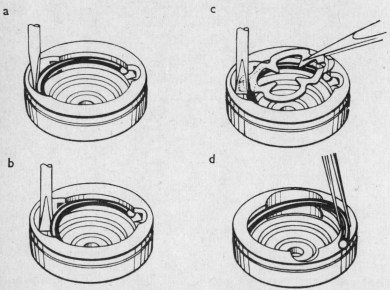

Fig. 60. Disassembling Kif Flector.

time-keeping and also excessive and irregular wear to the pivot bearings.

Spinning a balance staff between callipers and inspecting through an eye-glass will reveal any such damage.

It seems obvious, therefore, that some means of absorbing the shock is required. This in fact has been done and the principle employed is to use mobile pivot bearings that automatically take up a position of alignment under spring pressure (Fig. 58).

Two housings are employed. One is fitted into the balance cock and the other into the bottom plate. In these housings are assembled the jewels, the end-stones and the springs.

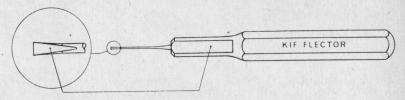

Fig. 61. Kif Flector special tool. The mark on the handle indicates the flat side of the point.

Shock will cause the balance to move from its normal position. The pivots will move the pivot bearings until the balance staff hits against the housings, and the housings and the balance staff will absorb the shock. The springs then return the assemblies to their positions of alignment.

Three of the more popular methods used have been selected and are described in the following paragraphs.

Kif Flector (Fig. 59). A pointed tool is inserted into the spring notch and the spring is turned in the direction of the arrow (*a*). The spring can now be hinged upward allowing removal of the end-stone and the setting jewel (*b*). The setting jewel and end-stone can now be cleaned, and the balance staff pivot is sufficiently exposed to be cleaned (*c*). Setting jewel and end-stone are replaced. The spring is hinged downward and the pointed tool is used to lock the spring by turning it in the reverse direction (*d*).

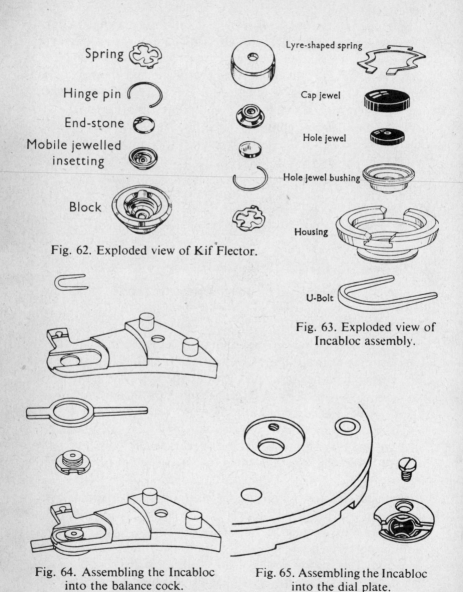

Spring

Hinge pin

End-stone

Mobile jewelled insetting

Block

Fig. 62. Exploded view of Kif Flector.

Lyre-shaped spring

Cap jewel

Hole jewel

Hole jewel bushing

Housing

U-Bolt

Fig. 63. Exploded view of Incabloc assembly.

Fig. 64. Assembling the Incabloc into the balance cock.

Fig. 65. Assembling the Incabloc into the dial plate.

The method for dismantling the Kif Flector is illustrated in figure 60.

The special tool (Fig. 61) is inserted behind the hinge pin (*a*). The tool is then rotated a quarter of a turn to push the hinge pin

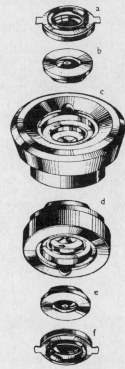

Fig. 66. Exploded view of Monorex. (a) Cap jewel in setting with cap jewel spring. (b) In-setting with jewel for balance. (c) Upper end-piece without fittings. (d) Lower end-piece without fittings. (e) In-setting with jewel for balance. (f) Cap jewel in setting with cap jewel spring.

inward (*b*). The spring is held by tweezers, unhooked from the hinge pin and removed (*c*). To remove the hinge pin from the block, one arm of the pin is first removed from its groove and the pin can then be lifted out (*d*).

Figure 62 shows the parts in their order of assembly.

Incabloc (Fig. 63). To dismantle, the spring must be swung clear. Press the ends towards the centre and raise the spring on its hinge. The jewel cap, the jewel and the jewel bushing can now be lifted from the housing.

In the case of the upper assembly the housing is secured to the balance cock by a U-bolt underneath (Fig. 64). The lower assembly housing is secured to the bottom plate by a screw (Fig. 65).

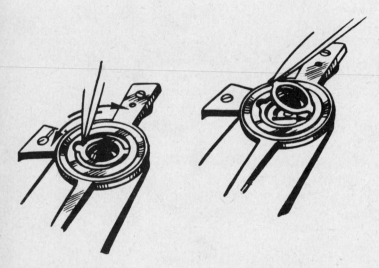

Fig. 67. Monorex shock absorber. Removing the jewel and bushing.

Monorex (Fig. 66). The end cap is removed as shown in figure 67. The jewel and bushing can then be lifted from the housing. Both upper and lower housings are press fits into the balance cock and bottom plate respectively.

One of the features of these three systems is the ability to expose both balance pivots without disturbing the balance movement. This enables the pivots to be cleaned as well as the shock absorber parts.

Oiling the pivots is accomplished by placing a drop of oil on the inner face of the end-stone and placing the end-stone and jewel together. They are then lowered into their housings and secured.

CHAPTER THIRTEEN

Cases

CASES are made of gold, silver, stainless steel or plated brass. Typical designs are illustrated in figures 68 and 69.

Pocket watch cases. Bezels are snapped on to the case middle. Domes are either hinged to the base of the middle or are fixtures and do not open, some cases have no domes. Backs are either hinged to the base of the middle or they are snapped on.

Wristwatch cases. Many are designed similar to pocket watches but the modern trend is to fit movements into two-piece cases as shown in figure 69. The bezel either fits over the back as illustrated or snaps on. Because the two-piece case has only one detachable part, the possibility of dirt and dust entering the case is reduced. This type of case is becoming increasingly popular among manufacturers.

The majority of American watch cases are fitted with screw-on bezels and backs. This is undoubtedly the best method of maintaining internal cleanliness.

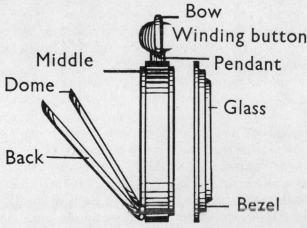

Fig. 68. Pocket watch case.

American manufacturers were the first to produce the gold-filled watch case.

A sheet of brass is sandwiched between thin gold sheet and all are brazed together. The metal is then rolled to the required thickness, pressed out and polished. The result is a brass case with a gold face inside and outside. These cases are very hard wearing.

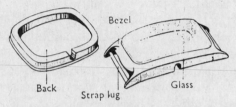

Fig. 69. Wristwatch case.

Opening a case. Before attempting to open a case it should be examined to make sure that the method of assembly is understood. A screwed-on back will be indicated by slots or by a series of flats cut in the edge to take a key or the edge may be knurled to enable the back to be gripped in the hand and unscrewed by hand pressure.

A screwed-on bezel will have a knurled edge.

Snaps can be identified by a shallow groove on the edge to provide entry for a knife blade, or the edge may have a lip to act as a lever.

If the snap has neither groove nor lip, a sharp knife blade is used inserted in the join.

When opening a good fitting case, great care must be exercised. The knife can slip and disfigure the finish. Damage can be caused to the edge of the snap. The join can be bruised by the incorrect use of the knife blade.

Most screw-on fittings can be removed without the use of tools. Place the watch face down in the palm of the left hand. Press down firmly with the other palm and at the same time turn in a counter-clockwise direction.

To remove a snap by means of a knife blade, first ensure that the blade is sharp. Next push the blade firmly and cleanly into the join. Now lean the blade over so that the cutting edge presses against the case middle allowing the blade face to push off the snap. Never twist the blade when it is in the join; the cutting edge will damage both the snap and the case middle.

Removal of the movement from the case is dealt with in Chapter 5.

Repairs. A bruise in the case middle can be pushed out from the inside by using a suitably rounded implement such as a piece of boxwood.

Bruises in domes and backs can be lightly tapped out with a wooden mallet whilst the dome or back is supported on a boxwood stake (Fig. 70). The stake should be covered with tissue paper to prevent marking the work.

Small bruises in the glass groove of a bezel can be pushed out

Fig. 70. Boxwood stake.

by using a small screwdriver from the inside. Considerable care must be exercised otherwise the glass will not fit correctly.

Side movement of hinges should be eliminated. In many cases the renewal of a hinge pin will be all that is required.

The old pin is carefully driven out a distance sufficient to enable the pliers to obtain a grip. The pin can then be pulled completely out.

Select a piece of steel rod of the correct diameter and cut it about $\frac{1}{8}$ in. longer than is required. Insert the pin so that an equal length is protruding at each end. These ends are carefully filed to blend with the shape of the case. File marks are removed by using the smoothest grade of emery cloth and then polishing.

Worn snaps can be tightened by inward burnishing of the edge with a wetted oval burnisher, or the snap on the case middle can be burnished outward.

A snap that is too tight can be loosened by removing a little of the metal all round the inner edge, finishing off with a burnisher.

Glasses. Some of the more popular designs are shown in figure 71. A loose glass can frequently be fixed in its bezel by applying some glue to the glass groove with the point of a pegwood stick. Place the glass in position, rotate it in the bezel to spread the glue and put aside to dry.

The edge of the glass groove should never be burnished inward to make a glass fit as this will damage the bezel.

Glasses are fitted by hand and to snap a good fitting glass into its bezel requires practice. Glasses should never be fitted with the bezel in position on the watch case.

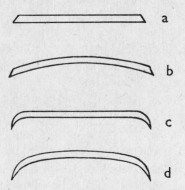

Fig. 71. Glasses. (a) Flat crystal. (b) Crystal. (c) Double lunette. (d) Lunette.

CHAPTER FOURTEEN

Magnetism

IT is not uncommon for watches to suffer from magnetism. The effect is erratic time-keeping, gaining at irregular intervals similar to that which is brought about by oil on the hairspring.

Much of the repairer's time can be spent chasing a watch fault whilst the real cause of magnetism remains undetected. Therefore, before we leave the overhaul of watches a few words about magnetism may save a few frustrating hours in the future.

For magnetism to have any noticeable effect on a watch it must be present in the hairspring. Other steel parts such as the keyless work, screws and mainspring are not likely to effect time-keeping if they themselves become magnetized.

In many modern watches the escapement is made from non-magnetic metals for this very reason. However, there are still large numbers of watches made with steel hairsprings that can be magnetized and it is with these movements that we are going to be concerned.

Remove the back of the case and with a pointed pegwood stick close the coils of the hairspring and see if they stick together when the pegwood is removed.

Now place a pocket compass on the balance cock. The compass needs to be the smallest obtainable for lightness and sensitivity of the needle. A very suitable type is the toy that is sometimes found in party novelties.

Position the compass centrally over the balance pivot. Set the balance in motion and note the effect on the compass. The needle may vibrate in harmony with the balance or it may make complete revolutions, but in either case the presence of magnetism has been established.

If the needle remains stationary, tap the compass in case the needle has stuck.

Lift the compass from the watch and take the watch away. If

the needle then moves it indicates that some other part of the watch is magnetized and was attracting and holding the compass needle in one position.

To check this, replace the watch under the compass in the same position as before and the compass needle should swing back.

Another check is to lift the compass just clear of the balance cock, slowly rotate the watch and see if the compass needle follows the movement of the watch.

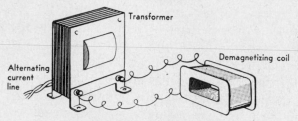

Fig. 72. Demagnetizer.

To demagnetize a piece of steel the part concerned is placed in the magnetic field set up around a coil of wire through which is passed an electric current. The part is first fully magnetized and then slowly withdrawn from the magnetic field.

The hairspring must be removed from the watch and wrapped in tissue paper in order to preserve its shape during the demagnetizing process.

If this were not done and the complete watch was put in the coil, or if the hairspring alone was put in the coil without protection, then when the electric current was switched on and the spring subjected to the full magnetic force the pull would far exceed the spring tension and the coils would become entangled.

The demagnetizer (Fig. 72), consists of a tightly wound coil on a hollow bobbin, the hole being large enough to accept watch parts with plenty of free space.

The coil is connected to an A.C. current through a transformer which provides the coil with a pressure of about six volts.

The hairspring is placed in the centre of the coil and the current is switched on. After a few seconds the hairspring is slowly withdrawn to arm's length and then the current is switched off.

It frequently happens that the first treatment of demagnetizing does not provide a complete cure. Therefore, when the hairspring is refitted to the watch and a further test is made, do not discard the possibility of magnetism still being present if the compass needle gives an indication.

CHAPTER FIFTEEN

The movement

As a time-piece the basic principle of a clock movement is the same as that of a watch. It has its motive power, train of wheels, escapement, and motion work with dial and hands.

Motive power is supplied either by a wound spring or by hanging weights.

From figure 24 we can see that the order of the wheels in a watch is: the barrel, the centre wheel, the third wheel, the fourth wheel and the escape wheel.

In a clock the barrel does not mesh with the centre wheel pinion. The order is: the barrel, the intermediate wheel, the centre wheel, the third wheel and the escape wheel.

The three most common types of clock escapements are the pin-pallet, the cylinder and the recoil. The pin-pallet is used mostly in the lower priced alarm clocks and watches and has been dealt with fully in Chapter 10.

The cylinder escapement is fitted to carriage clocks and is described in Chapter 20.

Whereas these two escapements are fitted with balance wheels, the recoil escapement, which is described in Chapter 16, is controlled by a pendulum.

When mechanisms such as strike and chime are added they tend to make the complete movement complicated. It is therefore better to make a study of these mechanisms separately. Part 3 has been written with this in mind.

The procedure for cleaning varies with some movements and in those cases it has been included in the chapter concerned, otherwise cleaning is the same as described in Chapter 16.

The technique of oiling has been fully described in Chapter 5 and applies equally well for clocks.

The dials of some movements are screwed to the front of the case. In others, the case itself is the dial.

Winding a pendulum movement is almost always done by inserting the key at the front. This is to avoid disturbing the clock and upsetting the balance. With other types of movement, where this precaution is unnecessary, winding is done at the back.

There are many methods of securing movements to cases, but whatever the method it will be obvious after inspection.

The remaining chapters in this section are devoted to different types of clock movements, it seems fitting therefore that in this chapter some mention should be made of clock cases.

Quite often a great deal of work is put into the overhaul of a movement which is then replaced in a case that contains enough dirt to stop the clock on sight.

Apart from the important task of cleaning the case interior, a few moments spent on the exterior often makes a big difference to its appearance.

Materials used for the manufacture of clock cases include wood, metal, stone, plastic and glass.

If the case is dirty give it a good cleaning. If it needs repairing, assuming the damage is not extensive, then a few minutes spent on a minor repair are worth while.

Clean the interior of the case. It is surprising how much dirt and dust can collect.

If the case is made of wood, examine ornamental pieces on the outside for security. Do the same with any strengthening blocks inside. Loose pieces can be glued into position.

Dust in carvings or scrolls can be removed by using a soft brush. Furniture cream will improve the appearance of polished wood.

Inspect bezel hinges and door hinges for security and condition.

Renew the screws if necessary. If the screws continue to turn when being tightened remove them and plug the holes with pieces of match-stick and replace the screws.

Other cases can be wiped with a clean soft cloth moistened in warm soapy water. Finish off with a soft dry rag.

Make sure the dials are clean. Lightly brush them with a watchmaker's soft cleaning brush. Enamel dials can be wiped with benzine.

Clean and polish the glass. Do nothing with lacquered brass-work other than rub it lightly with a soft cloth. Tarnished brass is best left alone.

CHAPTER SIXTEEN

Pendulum clocks

IN this chapter we are going to overhaul a simple pendulum clock of the lower priced domestic variety. It seems reasonable to say that these clocks are one of the most popular clocks in the home today.

Their low price is brought about by the high rate of mass production. Large quantities are manufactured in Germany, the United States and Great Britain.

The movements themselves are readily recognized. The plates and wheels are coated with lacquer and many of the parts are pressed out by machine. Providing they are kept clean, oiled and adjusted these clocks give excellent service over long periods.

In Chapter 15 we spoke about taking movements from their cases and removing dials and hands. Let us assume that this has been done and that we are now ready to proceed with the overhaul.

First of all we need a small but strong cardboard box on which to place the movement and so prevent the many protrusions from damage. Better still of course would be a home-made box of ply-wood cut to fit the movement.

Now let the mainspring down. If the movement is very old the chances are that the mainspring will not be contained in a barrel in which case a mainspring clamp will have to be fitted.

Place the key on the winding square and turn it in the direction of winding. This will throw the ratchet click out which must then be held in this position. Allow the tension of the spring to turn the key backwards until your hand can go no farther. Release the ratchet click back into the ratchet and then you may let go of the key. Repeat this process until the spring is completely unwound.

These springs are very powerful and could cause personal injury to hand or fingers if allowed to run down out of control. Apart from this, to do such a thing would most probably result in a broken spring. So take care and make sure that the movement is held firmly, that the key will not slip in your hand and that the click can be released immediately and at any time.

Remove the minute wheel collet, lift off the hour wheel and then remove the minute wheel.

Worn pivots and pivot holes have an effect on the depth of mesh between the wheels and their pinions and the locking of the escapement. Hold each arbor between the finger and thumb, or stout tweezers if the arbor is not accessible, as near the pivot as possible and check the amount of side-shake. Make a note of any holes that are considered to be in need of bushing. The best way of assessing whether or not the side-shake is excessive is to compare it with a movement of similar size that has had little or no wear.

Once having experienced the amount of side-shake in a new clock, subsequent inspections will prove less difficult.

Examine each arbor to make sure it has end-shake. There must be sufficient clearance to prevent the shoulders of the pivots binding against the plates.

Providing there is no excessive side-shake in the pallet arbor pivot holes and the escape wheel arbor pivot holes we can go on to check the recoil escapement, otherwise checking the escapement must wait until the worn pivot holes have been bushed.

Wind up the mainspring a few clicks, hold the movement in the left hand and operate the crutch with the right hand. Examine the faces of the pallets for wear. Observe through an eye-glass the action of the escape wheel teeth dropping on to the pallets and at the same time form an idea of the angle of swing the pendulum must make to release the escape wheel teeth.

It will be seen that deep locking will necessitate the pendulum passing through a large angle of swing which in turn demands more power. Remember also that if the pallet fails to release an escape wheel tooth just once, the clock will stop.

If on the other hand the depth of locking is too shallow mis-locking is likely to occur.

There are two types of pallets used, the strip pallet (Fig. 73, left), and the solid pallet fitted to the Garrard escapement (Fig. 73, right), both of which are provided with means of adjusting the depth.

In the case of the movement fitted with a strip pallet, the screw holes in the pallet cock are elongated and all that has to be done is slacken off the two screws, reposition the pallet cock and tighten the screws. It is a matter of trial and error but the correct position is quickly found.

Adjusting the depth of a solid pallet is done by altering the position of a screw on top of the back plate. Slacken off the two

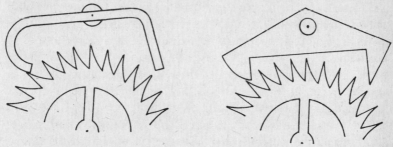

Fig. 73. Recoil escapements. Strip pallet (*left*) and solid pallet (*right*).

pallet cock screws, rotate the adjusting screw and tighten the pallet cock screws. By turning the adjusting screw in a clockwise direction the pallet cock is raised which reduces the pallet depth, and of course, to increase the depth the screw must be turned counter-clockwise.

Having checked and adjusted the pallet depth we have ensured that the escape wheel teeth will be properly locked. Now we can examine the amount of drop.

The drop of an escape wheel is the distance it travels from the release of a tooth from one pallet to the next arrest of a tooth by the other pallet.

To check the drop apply the same technique as was used when checking the pallet depth. In determining whether the amount of drop is correct, again the best advice is to compare it with a similar clock that has had little wear. Make a note of the adjustments that are needed then remove the pallet cock and lift the pallet arbor from the movement.

If the pallet is the strip type all that is needed is to bend the offending pallet or pallets. Bending the entry pallet outward will decrease the drop on that pallet, and bending the exit pallet outward will increase the drop on that pallet.

These strip pallets are quite soft and can be bent cold but as a precaution against cracking it is advisable to heat them first.

Now that the pallet is out of the movement this is a good opportunity to deal with worn pallet faces. Use a dead-smooth file to restore the faces to their original surface, smooth off with the finest emery stick and finish with a burnisher.

With the solid type of pallet the drop can be increased by reducing the entry faces. These pallets are too hard to file and a stone must therefore be used. An oilstone such as is used by a cabinet maker is best for this job. Make sure the stone is kept flat and that it follows the contour of the pallet.

The marks from the oilstone are then removed by an oilstone slip keeping the grain flowing with the pallet. Finish off with a very fine emery stick and then burnish.

Little can be done to decrease the drop in the solid type of pallet but this is not important. What is important is to avoid having too little drop. Such a condition can, after pivot holes are worn, lead to the pallets fouling the tops of the escape wheel teeth and stopping the clock.

One last word about checking the drop of the escape wheel. All escape wheel teeth must be checked with both entry and exit pallets. Any variation in drop on one side of the wheel to a position diametrically opposite will indicate the wheel being out of round.

With the pallet arbor out of the movement we are now ready to check the depth of mesh between the wheels and their pinions.

Gearing is a subject about which complete books have been written. It matters little whether we are interested in horology or civil engineering, the theory of gearing is the same. Correct depth of mesh, one of the many aspects of gearing, plays a very important part.

Figure 74, *a*, shows a wheel and pinion in correct mesh, *b* the mesh is too deep, and *c* it is not deep enough.

Our main concern is to ensure that the wheels and pinions in the clock movement do not suffer from too much depth of mesh causing harshness in operation. This can be checked by the sense

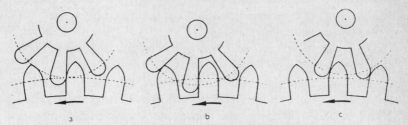

Fig. 74. Gear wheel and pinion in mesh. (a) Correct. (b) Too deep. (c) Too shallow.

of touch and is not as difficult as it may seem. After a little practice you will soon be able to detect those gears whose mesh is causing trouble.

We will commence with the third wheel and the escape wheel pinion. Hold the escape wheel by the tip of the first finger of the left hand and with the other hand rock the third wheel backward and forward. Although the movement is small any stiffness present can be felt.

Now press down lightly on the escape wheel arbor and slowly rotate the third wheel. Apart from your finger pressure causing slight resistance the action should be smooth and free from any roughness.

If tightness is experienced in one position but not in another the possible causes are:

        (*a*) damaged teeth
        (*b*) bent arbor
        (*c*) wheel out of round

If the teeth are broken or damaged beyond repair, new teeth can be fitted as explained in Chapter 6. If a tooth is bent towards an adjoining tooth it can be straightened by inserting the blade of a pocket knife between the two teeth and levering the bent tooth back into position. If the tooth is bruised it can be shaped with a thin smooth file and finished with a very smooth emery stick.

A bent arbor can be straightened cold. Find out first of all where the bend is by spinning the arbor and then place it on a flat steel stake and tap it straight with the peening end of a watchmaker's hammer.

If a wheel is out of round it can only be put right by renewing the wheel or by recutting the teeth with a topping tool.

The equipment for topping wheels is unlikely to be in the possession of a beginner, therefore this job will have to be sent away to be done. The same comments apply to bushing worn pivot holes. In both cases the repairer will require the front and back plates, pillar nuts and the wheels and pinions that are causing trouble.

To continue with the dismantling. Remove the pillar nuts and carefully lift off the back plate. Take out the escape wheel, third wheel, intermediate wheel and barrel. Hold the front plate in the palm of the left hand with the cannon pinion uppermost. Strike the centre arbor a sharp blow with a brass-faced hammer, this will free the cannon pinion allowing the centre wheel to be lifted out.

Now the barrel needs inspecting. Hold the square end of the barrel arbor between the jaws of the brass-faced pliers and try the end-shake. If there is none, place the barrel over a piece of tubing and gently tap the arbor and try the end-shake again. Repeat this until the end-shake is just perceptible.

It matters little which end of the arbor is tapped by the hammer but remember that the end-shake has an influence over the position of the barrel in the movement. Consideration should be given therefore to other parts in the immediate vicinity of the barrel in deciding which end of the arbor to hit.

Next, remove the barrel cover. This is done by levering it off with a screwdriver. Make sure it is marked before you take it off so that there is no difficulty in putting it back into its correct position.

Examine the condition of the spring. If it is clean and oiled with no signs of corrosion don't disturb it. If the spring is broken, or if it has to be removed for any other reason, then first remove the arbor. To do this it will be necessary to turn it back a little to unhook it from the eye of the spring.

Pull out the centre of the spring very carefully using the brass-faced pliers. As soon as enough spring to hold has been pulled out, the rest of the spring is allowed to uncoil itself from the barrel. This operation must be done slowly and carefully.

If the spring were allowed to fly out, it is doubtful whether it could be used again due to distortion.

This completes the dismantling of the movement.

It is a recognized fact that when two dissimilar metals move against each other, such as a steel pivot revolving in a brass plate, the harder of the two metals will show signs of wear first. This is because fine particles of dust embed themselves in the soft metal which then becomes an abrasive and cuts the hard metal.

It follows then that if pivot holes need bushing the pivots themselves will most certainly need polishing, and further to this, bushing cannot be carried out until the pivots have been polished because the bushes have to be drilled to suit the pivots.

When the work of bushing the plates is sent away the repairer will polish the pivots at the same time.

Having carried out all necessary repairs we are now ready to proceed with the cleaning. We will need a tin measuring about 6 in. × 4 in. and about $1\frac{1}{2}$ in. deep to hold the cleaning fluid. A $\frac{1}{2}$-in. paint brush makes a good brush for washing the parts.

We will also require a tapered strip of chamois leather, a piece of soft linen, some pith sticks, some pegwood and a folded newspaper.

Hold the parts over the cleaning fluid and well brush them with the wet brush, continually dipping the brush into the fluid. Make sure all the wheel teeth and pinion leaves are well brushed. Treat all the parts alike, both brass and steel, and lay them on the folded newspaper to drain. Use the linen cloth to dry them and to polish the arbors. Lightly brush the parts with a medium watch brush charged with chalk.

Shave a point on the end of a pegwood stick and peg out the bearing holes. Larger holes can be cleaned out by the strip of chamois leather. Clamp the wide end to the bench, thread the narrow end through the hole and run the plate up and down the strip.

Push the pith stick well on to each pivot and twist it round a few times.

If the mainspring has been washed, dry it with the piece of linen.

Finish off by brushing all the parts with a clean soft watch brush.

Now wind the mainspring into the barrel. Start off by hooking the outer end to the barrel wall and then slowly feed the spring in and at the same time rotate the barrel. When the spring is in do not attempt to push it down because of the possibility of causing damage. Tap the barrel on the bench two or three times, this will cause the spring to settle down to its correct position.

Now put the arbor in making sure that the eye of the spring is properly hooked and wrapped around the arbor.

Oil the edges of the mainspring and snap the cover on. Oil the arbor bearing hole in the cover and in the barrel. The movement is now ready for assembly.

Lay the front plate over the cardboard box and place the barrel and the wheels in position. Lower the back plate on to the movement and with a pair of tweezers manipulate the pivots into their respective holes

When all the pivots are in, the back plate should be resting on the pillars. The pillar nuts can then be put on finger tight.

Make sure that all the arbors have end-shake and then tighten down the back plate. Oil all the pivots in both plates.

Wind up the mainspring two or three clicks and check the wheel train for freedom.

Now fit the cannon pinion. Place the movement on a block of hard wood front plate uppermost. Push the cannon pinion on to the centre arbor and drive it home using a hollow punch and a hammer. Check the centre arbor to make sure it has end-shake. Again wind up the mainspring a few clicks to make certain everything is free.

Refit the motion work, the pallets and the pallet cock. Apply a spot of oil to each pallet and oil the pallet arbor pivots.

The movement is now ready to be fitted into its case. When this has been done, hang the pendulum and the movement should function.

Refit the hands making sure that when the hour hand points at one of the numerals the minute hand is immediately over the 12.

CHAPTER SEVENTEEN

Striking clocks

THE striking mechanism of a clock can best be understood if it is divided into two groups, the motive power and the release mechanism.

A typical striking clock is shown in Plate VII. The striking mechanism has been illustrated diagrammatically in figures 75 and 76.

It will be seen that when the strike mainspring is wound up, energy is transmitted to the fly wheel. On the face of the fly wheel is a pin. The upper end of the hour locking lever is bent so as to arrest the pin and prevent movement of the wheel train.

The cannon pinion cam lifts the hour warning lever which in turn raises the hour locking lever. As the hour locking lever rises:

(a) the fly wheel pin is released, the fly wheel rotates half of a turn and is again arrested by the hour warning lever.
This movement of the fly wheel is known as the 'warning'.
(b) The hour locking lever releases the gathering pallet which rotates a short distance during the half turn of the fly wheel.

The hour warning lever continues to rise and releases the fly wheel, the wheel train goes into operation and at the same time the rack falls on to the snail. The edge of the snail is shaped into twelve different steps, one for each hour. Whilst the wheels are turning the star hits against the hammer lifting pin causing the hammers to operate.

The gathering pallet revolves and the gathering pin engages the rack teeth lifting the rack one tooth at a time.

During the period when the pin is moving round to engage another tooth the rack is held by the hour locking lever engaging with the teeth.

Finally the rack is gathered up which allows the hour locking lever to fall and engage in the slot of the gathering pallet and at the same time to arrest the fly wheel from further movement.

Now to proceed with the overhaul. Let both mainsprings down. Remove the hammers, the back cock and the pallets.

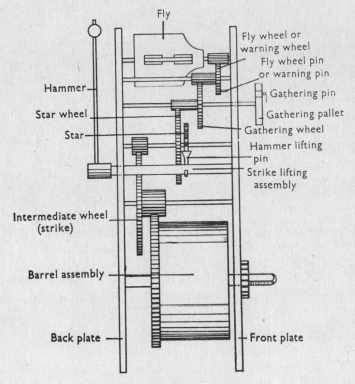

Fig. 75. Strike wheel train.

Unpin and remove the rack, the hour and minute wheels, the hour warning lever and the hour locking lever.

Inspect all pivot holes for wear and note those that require bushing. Place the movement, back plate uppermost, over a suitable box and unscrew and remove the back plate. Lift out the

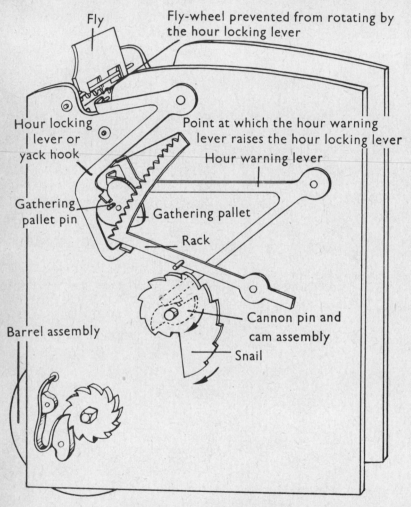

Fly

Fly-wheel prevented from rotating by the hour locking lever

Hour locking lever or yack hook

Point at which the hour warning lever raises the hour locking lever

Hour warning lever

Gathering pallet pin

Gathering pallet

Rack

Barrel assembly

Cannon pin and cam assembly

Snail

Fig. 76. Strike release mechanism.

barrels and mark one of them to ensure that they are refitted in their original positions, then lift out the wheels.

Remove the cannon pinion as described in Chapter 16 and treat the gathering pallet in the same way.

Carry out any necessary repairs and lay the parts out for cleaning. If the parts are lacquered it is sufficient to wash them and lightly chalk-brush them. Clean out the pivot holes.

We can now go ahead and assemble and oil the movement in the usual way and therefore we must consider the strike mechanism once again.

When all the parts have been positioned on the front plate and the back plate has been properly located, screw on two nuts finger tight diagonally opposite each other.

Turn the wheel train in the normal direction of rotation until the warning pin on the fly wheel is pressing against the hour locking lever. In this position the hammer lifting pin should be midway between two teeth of the star. If it is not, the back plate must be lifted and the star wheel disengaged and moved round. Replace the back plate and check again. When these two wheels have been correctly positioned the back plate can be fully tightened down with all four nuts.

With the hour locking lever in the downward position and arresting the movement of the fly wheel, the gathering pallet is fitted to its pivot so that the bent arm of the hour locking lever is engaged in the notch of the gathering pallet.

Operate the strike mechanism and note the position on the snail where the rack falls. If it does not drop on to the beginning of a step then the hour wheel will have to be moved round.

Complete the assembly of the movement and fit the dial. Turn the minute hand round until the strike operates. Fit the minute hand so that it points to 12 and the hour hand to point to the hour that had been struck.

CHAPTER EIGHTEEN

Chiming clocks

THE strike mechanism has been dealt with in Chapter 17 and so in this chapter we will discuss the chime mechanism only.

Diagrammatic sketches of the principle of the mechanism are shown in figures 77 and 78. There are many variations of this layout but the basic principle remains the same.

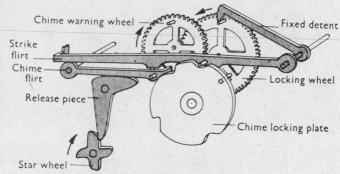

Fig. 77. Wheel train.

A star wheel with four arms is attached to the cannon pinion. The four arms represent the four quarters of an hour. One arm is longer than the other three and it is this arm that is in use on the hour.

As the star wheel moves round, the release piece is moved causing the chime flirt to pivot on its arbor. As the chime flirt moves it raises the strike flirt.

Secured to the arbor of the strike flirt is a detent with a hook end. This hook engages with a pin on the face of the locking wheel which prevents rotation of the chime train.

The strike flirt continues to rise until eventually the detent releases the locking wheel. This sets the chime train in motion but in the meantime the release piece has caused the chime flirt to lift higher and as the chime warning wheel moves round it is arrested by the chime flirt catching the pin on the chime warning wheel. The chime train is now stationary again.

This action takes place a few minutes before each quarter of an hour and, like the strike mechanism, is known as the 'warning'.

The chime train remains stationary until the star wheel has moved round to a position when the release piece has been lifted to the highest point of the star wheel arm. The release piece then drops and at that moment it is exactly a quarter of an hour.

The action of the release piece dropping causes the chime flirt

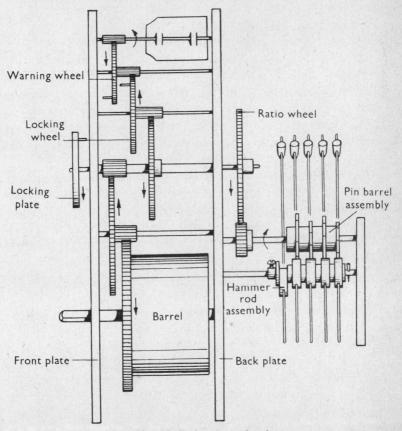

Fig. 78. Release mechanism.

to fall which in turn releases the chime warning wheel and allows the chime train to set in motion again. This time however there is no interruption and the train is free to rotate. In doing so the chime pin barrel is operated which in turn sets the hammer assembly in motion.

Immediately the chime flirt falls the chime locking plate lifts the strike flirt and holds the detent clear of the pin on the locking wheel. In this position the chime hammers will continue to function but in the meantime the locking plate is rotating. As it does so it brings a cut-away portion into line with the pin in the strike flirt and the strike flirt and detent drop. This causes the detent to arrest the locking wheel and chiming then stops.

The length of the cam faces on the locking plate are in the proportion of 1, 2, 3 and 4, each cam length representing that number of quarters of an hour. Thus on the hour the locking plate turns with its longest cam, representing four quarters, in contact with the pin in the strike flirt. This allows the chime train to rotate long enough for the hammers to strike the four quarters prior to the hour strike being released.

Just before the pin in the strike flirt drops into the cut-away portion of the locking plate, a pin in the locking plate engages the strike flirt and lifts it high enough for the strike rack hook to be released which releases the strike train. The strike flirt however catches the pin in the strike warning wheel and arrests the strike train until the full chime has been rung. The strike flirt then drops into the cut-away portion of the locking plate and the strike train is released without further interference.

Before dismantling the mechanism study it well. Cause it to operate until you know the sequence of operations sufficiently well to be able to assemble the parts without difficulty.

We will assume that the motion work and strike mechanism have been removed and that the springs have been let down.

First, remove the parts from the back plate. Take off the hammer rod assembly and the pin barrel assembly, then unscrew the ratio wheel. Unpin the strike lifting lever. Unscrew the pallet cock and remove the pallets.

Lay the movement on its back over a cardboard box and remove the release piece, the strike flirt, the chime flirt and the chime locking plate. The pillar nuts can now be removed and the front plate lifted off.

As the barrels and flys are lifted out mark them with pencil for identification: *C* (chime) *S* (strike).

In a movement of this kind there are many points to oil. Those which are common to all time-pieces have already been mentioned but with the addition of the chime mechanism there are many more.

To list them would be quite unnecessary as long as you remember that any two surfaces rubbing together need lubrication. This does not mean you are at liberty to exercise your skill with an oil can. The same strict discipline concerning oiling applies just as much with clocks as it does for watches.

When oiling the friction surfaces of the chime mechanism the smallest drop of oil will be enough.

CHAPTER NINETEEN

Grandfather clocks

A SPRING-WOUND clock derives its motive power from a coiled spring but a grandfather clock is powered by weights suspended from chains or lines of gut.

These weights provide a greater force than a clock spring, and consequently they are able to overcome resistances which in a spring-wound clock would cause the movement to stop.

This inevitably results in grandfather clocks functioning continuously for periods far greater than are usually achieved by spring-wound clocks.

Dirt, dust and lack of oil precipitate the rate of wear. It is a fact that these clocks collect inside their cases an extraordinary amount of dirt. With age, the oil partially evaporates and is partly absorbed by dust. Eventually no oil is left and the moving parts receive no lubrication.

The force of the weights continues to operate the clock and wear is therefore accelerated on all friction surfaces. Eventually this wear becomes so excessive that even the weights are unable to overcome the resultant resistance.

Examination of a movement in this condition will disclose badly worn pivots, elongated pivot holes, worn pallets and, quite often, rusty steel parts.

However, before commencing the dismantling of a grandfather clock, let us examine it in its case. Sometimes a fault can be found and rectified without removing the movement from its case which will set it in motion again. Make sure, of course, that the gut has not broken allowing a weight to fall to the bottom, or that the clock only needs winding.

The methods used to secure movements to their seat boards are often very poor and are a constant source of trouble. The seat board must be held rigid to the clock case and the clock movement must be held firmly to the seat board. The inertia in the swinging pendulum tries to pull the movement first one way and then the other and if the movement is allowed to remain insecure it may upset the beat.

Make sure the hands are not catching any part of the dial and are not fouling each other.

Inspect the gut lines for condition and freedom. The gut wears with usage and becomes frayed. Sometimes the strands foul in the movement. The lines should hang freely. If they are found to be rubbing against the seat board, then the slot in the seat board must be enlarged.

If when the preliminary inspection has been completed the fault has not been found, the movement must be removed from the case.

Remove the pendulum and the weights and withdraw the movement complete with its seat board. Remove the seat board from the movement. Pull out the collet pin and remove the hands. Withdraw the dial pins at the back and remove the dial.

If the condition of the movement is bad it will have to be brush-washed in an assembled condition before an inspection can be carried out. In severe cases the movement has to be dismantled and cleaned and then assembled for inspection.

There is one further examination to be made before dismantling the movement and that is of the escapement. Most grandfather clocks are fitted with the recoil type, so called because of the reverse movement made by the escape wheel each time a tooth has dropped from the pallets. It is the amount of drop from each pallet that has to be checked. To do this we must first make sure that there is no excessive shake in the pallet pivot holes and the escape wheel pivot holes. If these pivots are loose in their holes the drop of the escape wheel teeth will be affected.

Inspect also the pallets and the crutch for tightness on the arbor. Any looseness here will also effect the function of the escapement.

Details of how to check the drop of the escape wheel teeth will be found in Chapter 16.

Check all pivot holes for wear and make a note of those that need bushing.

The movement may now be taken to pieces, examined and notes made of any repairs that are needed. The parts are then laid out in readiness for cleaning.

There are a number of cleaning fluids that can be made up, all of which do the job very well. A cheap and well tried one is to mix four ounces of soft soap with two pints of hot water and add two teaspoonsful of liquid ammonia. Any big hardware store will supply the ingredients.

When making up the solution stir well to ensure that the soft soap is completely dissolved. Pour in the ammonia last.

Allow the solution to cool. Place the components in a tray and pour the solution over them until they are covered and leave them to soak for twelve hours.

The parts are then lifted out, brush-washed in clean warm water and dried in warm boxwood dust in a metal container held over a spirit flame.

Steel parts that have become rusty can now be dealt with. Flat pieces are best laid on a flat wood block and cleaned with the finest grade of emery stick. The stick must be held flat and square to the work and the grain must follow the length or contour of the part being cleaned. Round parts and flat circular parts are best held in the turns and cleaned whilst being spun.

If the rust on screws cannot be removed easily, then they are best renewed.

Rusty pinions can be cleaned by using a folded piece of very fine emery paper smeared with oil.

We can now go ahead and chalk-brush all the parts. A cleaning brush with medium bristles will do for this operation. The pivot holes are next to receive attention. The large holes are best cleaned with chamois leather. Cut a tapered length and secure the wide end in the padded jaws of a vice. Feed the narrow end through the pivot hole to be cleaned and run the plate up and down the chamois strip a few times.

The smaller holes are cleaned with pegwood. Insert a tapered pegwood stick in the hole, rotate the stick, withdraw it and shave off the dirt removed from the hole. Repeat this until all the dirt has been removed. After the pivot holes, make sure that all the oil sinks are clean.

Now that all the parts have been cleaned and repaired the

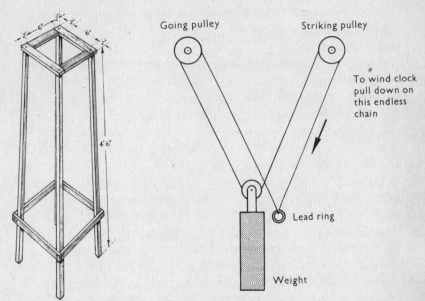

Fig. 79. Wooden stand.   Fig. 80. Method of hanging weight.

movement is ready to be assembled and oiled. Take care to ensure that all working parts have freedom of movement.

Secure the movement to its seat board and place it on a stand. A simple wooden stand is illustrated in figure 79. Check the level of the seat board in both directions with a spirit level. Adjustments can be made to bring the seat board horizontal by packing small pieces of thin card underneath.

Some clocks have only one weight which is suspended on an endless chain. Figure 80 shows the correct method of hanging the weight.

Fit the pendulum, hang the weights and partly wind up the

barrels. The beat has now to be checked. Move the pendulum slowly to one side and watch the escape wheel. Immediately a tooth drops note the angle of the pendulum. Move the pendulum in the reverse direction and note the angle it makes when the next tooth drops. The two angles should be the same.

If it is found necessary to move the pendulum further one way than the other the crutch has to be bent slightly in the direction of the greater angle.

Continue to adjust until the escape wheel teeth drop with the pendulum swinging equally each side of the perpendicular centre.

It follows that if the seat board is set horizonally in its case, the beat of the movement will remain correct.

The pendulum should be quite steady during its swing. If the pendulum bob can be seen to turn, watch the movement of the pendulum suspension spring in the slot of the suspension bridge. The spring should be a tight fit. If the slot is too wide, this will allow the spring to twist whilst the pendulum is in motion. If no fault is found here, remove the spring and examine it. These springs sometimes fracture at the point of flexing.

Check the striking mechanism for correct functioning. Fit the dial and the hands and the movement is ready to be fitted in its case.

Support the weights by holding the lines, release the clicks and slowly lower the weights until all the line has been unwound from the barrel drums. Remove the weights, lift off the pendulum and coil the lines neatly.

Before fitting the movement into the case, you will do well to give the case some attention. With the aid of a plumb-bob and line check the case to see if it is upright, then with a spirit level check to see whether or not the case is horizontal.

There are numerous ways of adjustment. A lot depends on where the adjustment is needed and how much. Packing pieces can be fitted underneath the case. Blocks can be inserted between the clock and the wall to push the top of the case forward.

When these adjustments have been carried out lift the movement into the case and check the seat board for horizontal.

Pack it underneath where necessary and then secure firmly to the case. Hang the pendulum and the weights and wind the weights up. Make sure that the lines are correct in the barrel drum grooves. Push the pendulum to one side and then release it; the movement should now function.

CHAPTER TWENTY

Carriage clocks

WHEN carriage clocks were made they were intended as travelling clocks. The movement is encased in glass panels mounted in a gilded frame. The clocks were supplied with a leather-covered felt-lined wooden travelling case which was cut away at the front to expose the dial. On top of the case was fitted a lid with a spring-loaded snap fastener. Raising the lid exposed the escapement which operated on top of the movement. The lid also

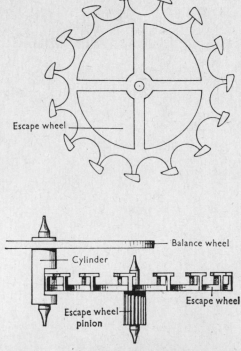

Fig. 81. The cylinder escapement.

provided access to the clock carrying handle. To gain access to the hand-set and to wind the mainspring it was necessary to remove the clock from its travelling case and open the glass panelled door at the back. These clocks were made with either a lever escapement or a cylinder escapement. Because the lever escapement is dealt with in Chapter 10 we will concern ourselves here only with the cylinder escapement as illustrated in figure 81.

The escapement consists of a balance, a cylinder and an escape wheel.

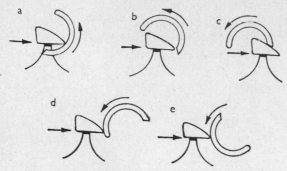

Fig. 82. Cycle of movement.

The balance wheel fits over the upper end of the cylinder and the teeth of the escape wheel pass through the cut-away portion at the lower end of the cylinder.

By referring to figure 82 we can see the cycle of movement:

(a) The cylinder has just reversed direction and is commencing a new swing.
(b) The tooth of the escape wheel is about to leave the cylinder.
(c) The tooth is giving impulse to the balance.
(d) The cylinder has moved round and arrested the next tooth.
(e) The cylinder completes its swing.

Unlike the balance of the lever escapement, the balance of the cylinder escapement is never free of the escape wheel in that the escape wheel teeth are at all times in contact with the roller.

The escapement is fitted to a platform which is mounted across the two movement plates. This is known as a horizontal escapement, and provided the clock remains horizontal the poise of the balance remains unaffected.

Because the escapement is positioned horizontally and the wheel train operates in a vertical plane it is necessary to provide a means of changing the direction of the motive power. This is achieved by a contrate wheel, the teeth of which are cut at right-angles to the rim of the wheel and engage with the leaves of the escape wheel pinion.

This design is not good gearing practice and it frequently causes trouble. The meshing of the contrate wheel with the escape wheel pinion is critical and care must be exercised when making adjustments.

The movement is held to the base of the case by screws inserted up through the bottom. Removal of these screws permits the withdrawal of the movement from the case.

The clock is dismantled in the normal way and made ready for cleaning. Special attention must be paid to the steel parts that are gilded, and to lacquered brass. Harsh brushing or the use of an abrasive will destroy the finish.

Wash the parts in the cleaning fluid and use only a brush with soft bristle after which all handling should be carried out using tissue paper or a soft cloth.

The inside of the cylinder must be scrupulously clean. Cleaning is best done with a tapered pegwood stick.

Any signs of wear on the cylinder lips are best eliminated by fitting a replacement cylinder. The function of the escapement is improved by burnishing the edges of the cylinder and the impulse faces of the escape wheel teeth. The teeth are then cleaned with a pith stick.

In the final stages of assembly check the end-shake of the contrate wheel; it should be as little as possible.

Means of adjustment is provided in the form of a screw at the back of the movement. The screw is held in position friction-tight by a slotted block. If the screw has become loose, the block must be removed and the slot closed. Now fit the platform and tighten the screws so that they only just nip and then assemble the escape wheel to the platform.

The depth of mesh between the contrate wheel and the escape wheel pinion can now be checked. If the escape wheel pivot is

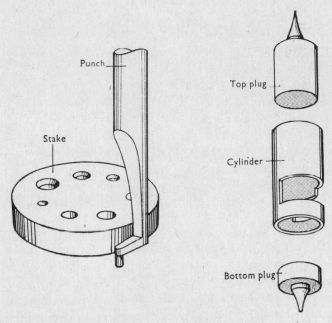

Fig. 83. Tools for removing cylinder plugs.

held by a piece of pegwood and the contrate wheel is rocked it is possible with the aid of an eye-glass to see the depth of mesh.

An adjustment either way can be made by tapping the end of the platform in the desired direction.

The screw holes in the platform are slightly oversize to permit such adjustment to be made. When the depth of mesh is correct the platform screws can be tightened.

By winding up the mainspring one or two clicks and setting the wheels in motion an impression is obtained as to whether or not the depth of mesh needs any further adjustment. A noisy contrate wheel is a definite indication of the mesh being too deep.

If the leaves of the escape wheel pinion are worn, a new surface can be offered to the contrate wheel by raising the platform. Four thin collars of equal thickness placed one under each corner will be sufficient.

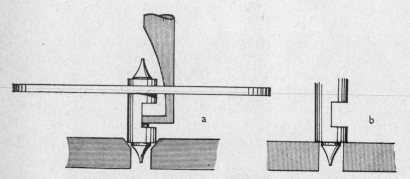

Fig. 84. Punching out bottom plug.

Now fit the balance and wind the mainspring up a little more. Before carrying out any further checks on the escapement it is necessary to find out if the height of the cylinder is correct in relation to the escape wheel. From figure 82 it will be seen that the escape wheel moves only when imparting an impulse to the cylinder. If, however, the cylinder is too low, allowing the edge of the cut-away portion to hit against the rim of the escape wheel, a backward movement of the escape wheel can be seen.

To check this, hold the balance wheel in the tweezers and slowly move it backward and forward through the normal angle of swing and observe the reaction of the escape wheel.

If the cylinder is too low it can be raised by driving the bottom plug out a little, and then driving the top plug in by the same amount. The tools for this operation are illustrated in figure 83. and the method is shown in figure 84. The stake illustrated has five plain holes and three countersunk holes all of different sizes.

A countersunk hole is selected for size, the cylinder is placed in the hole and the plug is partially driven out by the punch. To completely remove the plug, the cylinder is transferred to a plain hole the diameter of which will allow the plug to pass through.

We are now ready to check the locking of the escape wheel teeth. Once again hold the balance wheel in the tweezers and slowly lead it round until a tooth drops. At this point note the position of the balance wheel. Now lead the balance wheel round in the reverse direction and watch the escape wheel very closely. Immediately the escape wheel starts to move stop the balance wheel and note its new position. The total distance travelled by the balance wheel should be 5 degrees.

To assist this check, some manufacturers punch three dots on the face of the balance wheel, the two outside ones being 5 degrees from the middle one.

Immediately a tooth drops, a mark can be made on the plate underneath the centre dot. Then when the balance wheel is led round to its second position it can be seen whether or not the 5 degrees is being achieved. If it is less, then the locking is shallow and the cylinder needs to be brought closer to the escape wheel.

To do this, the platform is removed from the movement and the chariot screw underneath is loosened. The chariot is then moved closer to the escape wheel and the screw tightened.

This check must be carried out to all the teeth and with both inlet and outlet lips of the cylinder.

The next check is the shake of the teeth. Lead the balance wheel into the position shown in figure 85, a. By holding the escape wheel in tweezers try the shake between the points where the toe of the trailing tooth touches the inlet lip and the heel of the leading tooth touches the exit lip.

The balance is now moved into the position shown in figure 85, b. The internal shake can now be tried.

If the inside shake is the greater it is an indication that the cylinder is too large, and if the greater movement occurs on the outside of the cylinder then the cylinder is too small. The design of the escapement will, however, permit a generous tolerance.

Only if the inside shake is tight with a very loose outside shake is it necessary to take rectifying action. The remedy is to renew the cylinder by fitting one with a larger diameter.

Set in the rim of the balance wheel is a small banking pin which is arrested by a larger pin fitted in the back of the balance cock. The purpose of this pin is to prevent the balance wheel from swinging through an angle greater than 180 degrees in either direction. If the balance wheel exceeded this amount of swing it would foul the escape wheel teeth and the movement would stop.

If the balance wheel pin is missing it should be replaced. If it is worn, broken or bent it should be renewed otherwise there is the danger of it trying to pass the pin in the balance cock and becoming fast.

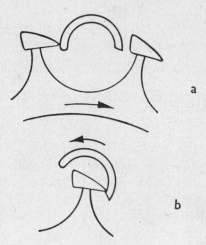

Fig. 85. Checking shake of escape wheel teeth.

CHAPTER TWENTY-ONE

Cuckoo clocks

THE popularity of cuckoo clocks has returned and today there is a thriving market in Europe catering for the tourist trade.

These clocks are smaller and less complicated than the original cuckoo clocks and they are less expensive. The originals are fast becoming antiques and are being sought after by dealers.

It is the older type of clock we are going to discuss because it is thought that an understanding of this mechanism will

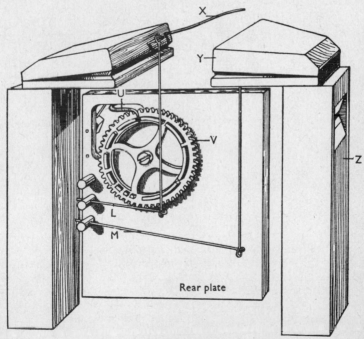

Fig. 86. Rear view of bellows (see Fig. 87 for key).

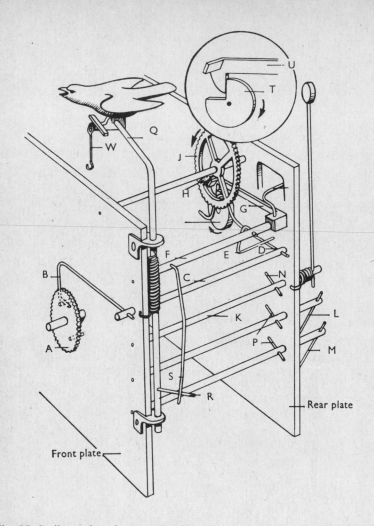

Fig. 87. Strike and cuckoo mechanism without wheel train. A. Cannon pinion. B. Lifting piece. C. Lifting arbor. D. Raising wire. E. Engaging pin. F. Locking arbor. G. Fly wheel locking piece. H. Fly wheel pin. J. Fly wheel. K. Hammer arbor. L. Bellows wire (short). M. Bellows wire (long). N. Hammer pin. O. Pin wheel. P. Bellows pins. Q. Cuckoo arbor. R. Cuckoo arbor pin. S. Cuckoo operating pin. T. Locking wheel. U. Count wheel locking piece. V. Count wheel. W. Door wire. X. Tip wire. Y. Bellows. Z. Pipe.

undoubtedly result in a working knowledge of a more simple movement.

Some clocks are weight driven and others are spring-wound but the method of motive power has no influence over the striking and cuckoo mechanisms which is all we are interested in here.

The case is made of wood with a carved front, the whole being roughly made. The cuckoo door is mounted above the dial and two inspection doors are fitted one in each side of the case.

Cut in the top of each side is an opening through which can be heard the sound produced by the pipes.

At the back of the case is a full-sized panel that can be lifted off leaving an aperture large enough for the withdrawal of the movement. On the inner face of the rear panel is mounted the gong.

Figure 86 shows the back of the movement with the pipes and bellows positioned on each side. The slot at the top of each pipe is opposite the aperture cut in each side of the case.

Before taking the movement from its case let us study the illustrations and find out how the strike and cuckoo mechanisms operate.

The key in figure 87 applies also to figures 86 and 88.

Attached to the rear of the cannon pinion A are two pins diametrically opposite each other. These pins operate the lifting piece B on the hour and the half-hour.

The lifting piece B is attached to one end of the lifting arbor C and at the other end the raising wire D is attached. It follows then that any movement of lifting piece B is transmitted to raising wire D.

Placed above the lifting arbor C is the locking arbor F. Attached to this arbor is the engaging pin E, the fly wheel locking piece G, the count wheel locking piece U, and the cuckoo operating pin S all of which move together.

When the raising wire D is lifted it contacts the engaging pin E and turns the locking arbor F.

Until now the hook at the end of the fly-wheel locking piece G has been resting in the recess of the locking wheel *T* and the tongue at the end of the locking piece *G* has been arresting the movement of the fly wheel pin *H*.

With the turning of the locking arbor *F* the lifting piece *G* is going to be raised and in doing so the pin *H* is released allowing the wheel train to function.

Almost immediately after, the pin *H* is again arrested this time by the crook in the raising wire *D*. This brief movement of the wheel train is known as the 'warning'.

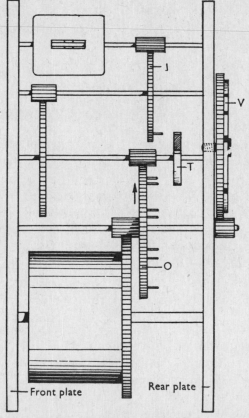

Fig. 88. Wheel train (See Fig. 87 for key).

In the meantime the cannon pinion has continued to rotate and now the lifting pin moves clear of the lifting piece *B* which drops. It follows that the raising wire *D* must also drop and in doing so the fly wheel pin *H* is again released and the train of wheels goes into motion.

Around the edge of the pin wheel *O* are seven pins spaced at equal distances. As the wheel rotates the pins trip the bellows pins *P* causing the wires *L* and *M* to operate the bellows. The same pins then carry on to trip the hammer pin *N* causing the hammer to operate. Due to the angular movements of the hammer and the bellows wires, the hammer strikes first followed by the long bellows wire *M* sounding 'CUC' and finishing with the short bellows wire *L* sounding 'KOO'.

When the raising wire *D* dropped, the locking arbor *F* swung round and caused the hook of the locking piece *G* to drop into the recess in the locking wheel *T* but the wheel train was in motion and therefore the cam action of the locking wheel *T* raised the locking piece *G* again.

At the same time that the locking piece *G* is being raised the count wheel locking piece *U* is being lifted from one of the slots cut in the rim of the count wheel *V* and the pinion on the end of the pin wheel arbor is causing the count wheel *V* to slowly rotate.

The result of this action is that when the recess of the locking wheel *T* again comes opposite the hook end of the locking piece *G*, the hook end is prevented from dropping into the recess by the locking piece *U* riding on the rim of the count wheel *V*.

The rim of the count wheel *V* is cut into eleven segments each longer than its neighbour. The strike and cuckoo action will continue to function for so long as the mechanism is kept in operation by the locking piece *U* riding on a segment of the count wheel *V*.

When the count wheel *V* has moved round to bring the next slot opposite the locking piece *U*, the locking piece *U* will drop causing the locking piece *G* to drop and the train of wheels will then be arrested by the fly wheel pin *H* coming into contact with the tongue on the end of locking piece G.

When the locking arbor F turns, the lower end of the cuckoo operating pin S moves inward, presses against the cuckoo arbor pin R and turns the cuckoo arbor Q. This swings the cuckoo forward and opens the door. When the cycle is complete, the cuckoo returns under spring pressure and pulls the door shut by means of the door wire W.

On top of the left-hand bellows is a tip wire X. When the bellows operate, the tip wire is lifted and it contacts the underside of the cuckoo lifting the cuckoo's tail up and lowering the head. This rocking action causes the wings to spread and the beak to open.

To take the movement out of its case the pipes and bellows must first be removed. Lift off the pendulum or weights. Disconnect the bellows wires. The pipes are secured by screws passing through the sides of the clock case. Remove these screws and the pipes can then be withdrawn complete with the bellows which are glued to the top of the pipes. Put them in a safe place away from possible harm.

Open the cuckoo door and disconnect the wire linking the door to the cuckoo, and then remove the hands.

All that now remains is to remove the screws holding the movement to the case and withdraw the movement.

The movement is dismantled and assembled as previously described in Chapter 18. Cleaning is carried out by washing the parts in a bath of cleaning fluid and then finishing with the chalk-brushing method. The illustrations will assist in the assembly of the strike and cuckoo mechanisms. To check their action after assembly place the tip of the forefinger on the teeth of the fly-wheel and allow the wheels in the train to rotate slowly.

To repair a broken pipe or even to make a new one is quite simple. The wood can be cut from a cigar box, shaped and glued. Bind the pieces together with strands of wool until the glue is set. When making a new one there are two things that require careful attention; the width of the air gap between the bellows and pipes and the internal depth of the pipes. If either of these is incorrect the tone or note of the pipe concerned will be affected.

If the bellows are leaking the kid must be removed and a new piece fitted. Fold the new piece using the original as a pattern. Press the new piece so that the creases become permanent and then glue into position.

The following amusing example of horological English was found on the back of an old clock:

'Indication of the manner to serve Cuckoo Clocks.

At first one draw out the bellows with much circumspection the paper and the cramps. Further it is necessary to remove the iron-wire or the packthread which is fixed on the chain. If the clock no strike correct it is necessary to open the small door of the right side and to press of the visible wire. If the clock do hang a little oblique, it is necessary, to bend the wire-sling (which is visible under the case of clock and as hang the pendulum) a little of the same side as hang oblique the clock.

By cuckoo clocks with sounding springs one draw the paper much cautious out the springs. Moreover it is to remove the fixed iron-wire upper the small door on the front of clock.

Cuckoo and Quail Clocks. As because the strike correct the Quail, it is necessary to open the small door of the left side from case of clock and to press of the visible iron-wire. If the clock march too quick one push the sheave from pendulum more upwards, if the clock march too slow one push and sheave more downwards.'

CHAPTER TWENTY-TWO

French clocks

DUE to the nature of French clocks being show-pieces as well as first-class time-pieces, extra care and trouble is expended during cleaning.

Fundamentally the sequence is clean, wash, dry and polish.

The recommended cleaning agent is a liquid metal polish. If this is applied with a stiff watch brush, a very fine finish can be achieved.

Shake some metal polish into a shallow tin and dip the end of the brush into it.

The part to be cleaned is held in the hand in a piece of soft cloth such as an old handkerchief. Brush inside the bars of the wheels with a side motion so that the grain will follow the contour of the aperture.

Take extreme care not to knock the pivots with the brush because apart from being very fine they are also quite hard and easily snap off.

When brushing the teeth make sure that the brush movement is made square with the wheel to ensure that the bristles reach the base of the teeth. Be very meticulous with the gearing and round the wheels two or three times.

When brushing the sides keep the brush movement always in the same direction. Never cross the grain.

Plates and other flat pieces are similarly treated, the grain always flowing with the length of the part.

To clean the oil sinks, place a hand-drill in the vice and secure in the chuck a short length of pegwood shaped like a flat drill. Cut a small square of chamois leather, smear it with a spot of metal polish and place it over the oil sink. Bring the work up to the pegwood stick and rotate the hand-drill at a good speed causing the chamois leather to spin in the sink.

When all the brush work is finished, take a leather buff

(supplied for the purpose) smear a very little metal polish over its surface and buff all the plates, bridge pieces, etc., retaining the same direction of grain.

All these parts now have to be washed in a cleaning fluid. Lay them in a pan of suitable size, pour the fluid over them and with a $\frac{1}{2}$-in. varnish brush remove every trace of the metal polish.

As each part becomes washed lift it out of the tray and lay it on folded newspaper to absorb the surplus fluid. The parts are then dried in a linen cloth.

Polishing is carried out with a soft watch brush charged with billiard chalk. All the parts are brushed as before, but of course, not so energetically, and for this operation they are held in tissue paper.

The oil sinks are polished using the same method by which they were cleaned only this time dry chamois leather is used.

Finally, all pivot holes, threaded and plain holes and pinion leaves are cleaned out and rubbed over with a pointed pegwood stick. When cleaning out the pivot holes the pegwood stick is inserted at both ends and rotated. Any dirt on the stick after withdrawal is shaved off and the stick is inserted again. This process continues until all traces of dirt have been removed.

As each part becomes finished it is put in a box away from dust.

Blued steel screws that have become marked should be restored to their original condition. This is best done by spinning the screw in a lathe. Remove the blueing and the marks with a dead-smooth file, polish with a piece of very fine emery paper and finish with a flat burnisher.

The screw head must not be touched by hand. Clean out the screwdriver slot and remove the screw from the chuck with tweezers.

There are two methods of blueing and preservation. One is to place the screws on a blueing pan (Fig. 11) and move it about in a spirit flame until the screw heads turn the right colour and then tip them into a tray containing thin oil. The screws are then washed in cleaning fluid.

The other method is to place them, head uppermost, in a tray of hot silver sand supported over a flame. When the head has turned blue the screw is lifted out with an old pair of tweezers and given one coat of colourless lacquer with a $\frac{1}{2}$-in. flat lacquer brush.

CHAPTER TWENTY-THREE

Alarm clocks

THERE is a wide range of alarm clocks on the market especially in the lower-priced models, and different methods of casing are employed.

Before dismantling an alarm clock it will be necessary to study the case to understand the design and method of assembly.

It is usual for the back of the case to be marked with arrows to indicate the direction of turning the winding keys. The reverse direction will unscrew them.

If the winding-click is broken any attempt to unscrew the key will only result in the key and arbor turning together. To overcome this, some clocks have a slot machined in the outer end of the spring arbor. A screwdriver is held in the slot and the arbor held stationary while the key is unscrewed.

The hand-set button is usually a push-on fit and only needs a straight pull to remove it. The alarm-set button is removed by unscrewing it against the direction of setting the alarm.

Sometimes the threads of the feet pass through the case and into the movement plate preventing the movement from being withdrawn. Unscrew the feet by wrapping them in two or three thicknesses of cloth and gripping them in pliers.

The round case is the most common type in the lower-priced models. The bezel either pushes into the front of the case, or it forms part of the case and the back pushes in from the rear. Sometimes the back is held in position by three or four screws.

Having removed the external parts and opened the case, the movement is now ready to be lifted out complete with dial and hands.

The minute hand and the alarm hand can be removed by using the lever shown in figure 10. Now bend back two of the tabs holding the dial to the movement and remove the dial complete with the hour hand and hour wheel.

159

Take care not to bend the dial tabs any more than is absolutely necessary because they have to be bent again during assembly, and too much bending will cause them to break off.

Cut off a short length of pegwood whose diameter will just fit inside the hour hand. Support the dial in the left hand and place the tip of the thumb and fingers between the wheel and the dial. Push the pegwood punch into the hour hand and gently tap the punch until the hour wheel and the hour hand separate.

It is unlikely that any work will have to be done on the dial or hands and so, when they are free, put them in a safe place away from the possibility of damage. If the hands and dial numerals have been coated with luminous paint, extra care must be taken during handling to prevent the paint from cracking.

Pull out the alarm wheel pin and remove the alarm wheel.

Before dismantling the movement, the escapement must be examined. The majority of the lower-priced alarm clocks are fitted with the pin-pallet type of escapement and it is therefore necessary to read Chapter 10 in conjunction with this chapter.

If when checking the shake of the lever against the balance staff it is found that the shake is unequal, the lever can be bent sideways without removing it from the movement. A suitable tool for this purpose can be easily made and is illustrated in Figure 89.

Having completed the inspection of the escapement, the balance may now be removed. Pull out the balance spring pin from its stud and guide the spring out of the stud and index pins. Slacken off the upper balance bearing screw and remove the balance from the movement.

With a pair of tweezers check the end-shake and side-shake of the wheel train pivots and the pallet pivots. Any pivot holes that have become enlarged should be noted. They can be dealt with after the movement has been dismantled.

Fully wind the mainspring and the alarm spring and hold them in this position by fitting spring clamps.

Place a piece of pegwood through the escape wheel and operate the lever continuously until the pegwood prevents any further rotation of the escape wheel.

Slacken the plate nuts just enough to lift the lever from the movement. Tighten the nuts, remove the pegwood from the escape wheel and allow the wheel train to rotate until the spring clamps prevent any further movement of the springs. When the wheel train comes to rest, the plate may be removed and the wheels lifted out.

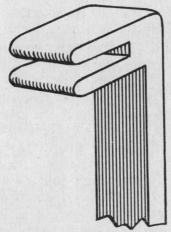

Fig. 89. Tool for bending lever.

In many alarms the cannon pinion is a driving fit to the centre arbor, and, once removed, may be always loose on the arbor. It is better therefore to leave this undisturbed and clean the pivot hole below the pinion by soaking and careful brushing.

Having dismantled the movement and laid out the parts separately we can now continue with the inspection.

Examine the balance staff ends for wear. Usually it will be found that the conical faces have worn away and produced a ridge. To rectify this the balance spring must first be removed. Mark the balance wheel to show the position of the collet. Make a tool similar in shape to a watch screwdriver but with a longer taper and push the blade into the slot of the spring collet.

A slight twist and pull made in one movement will remove the collet and spring from the balance staff.

Fit the staff in the turns and with an Arkansas stone restore the conical face making sure that the original angle is not destroyed. Burnish the new face and the point. Treat the other end of the staff in the same way.

Examine the pallet pins for wear. If they need renewing the old ones can be tapped out over a graduated stake. New pins can be

Worn pivot hole be-
fore reaming.

Pivot hole after ream-
ing, showing entry of
tapered bush.

Bush pressed into
position ready to be
filed to length.

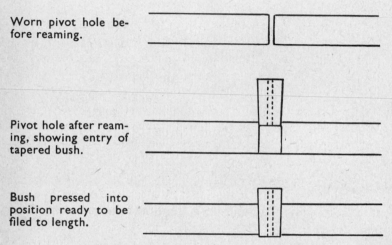

Fig. 90. Fitting Bush to worn pivot hole.

made from sewing needles. Cut off a suitable length and tap the new pin in position from the underside of the lever.

Enlarged pivot holes can now be dealt with by fitting new bushes.

Bushes are supplied ready made. The pivot hole is drilled and the outside diameter is machine tapered. Each bush is made oversize in length.

A convenient method of purchasing is to obtain from any materials supplier a box containing bushes of assorted sizes.

Select a suitable bush and ream out the worn pivot hole from the back of the plate until the diameter of the hole will just allow the small end of the bush to enter. The bush can then be pressed or tapped into position.

When the bush is in tight both ends must be carefully filed flush with the plate.

Finally the pivot hole is enlarged by reaming, to suit the size of the pivot.

Examine the wheel train for broken teeth. New teeth can be fitted as described in Chapter 6, but in place of brass pin-wire, strip brass will probably be needed.

Examine the wheel pivots and if wear has taken place the pivots must be polished and burnished in the turns.

Provided no other repairs have to be carried out we are now ready to proceed with the cleaning. A tray will be required for the cleaning fluid about two inches deep and large enough to take the back plate. We will require some tissue paper, two cleaning brushes, soft and medium, a ½-in. flat paint brush, pegwood, pith, chalk, a piece of clean soft linen such as an old handkerchief and some cleaning fluid.

Place one of the plates in the tray and pour the cleaning fluid over it until it is covered. Brush it with the ½-in. brush and when it is thoroughly clean remove it from the cleaning fluid and dry it in the cloth. Treat the other plate in the same way.

The balance spring is held in the tweezers and gently washed in the cleaning fluid by moving it about. When it is clean, lift it out and place it between two pieces of tissue paper changing the paper each time it becomes saturated.

Do not immerse the mainspring and the alarm spring in the cleaning fluid because, in their wound condition, they will not dry.

Wash, brush and dry all the remaining components in the same way as was done for the back plate.

Take the cleaning brush with the medium bristles and charge it with chalk. Rub it briskly over the faces of both plates. This will produce a semi-polish which is less likely to attract dust.

Treat the wheels in the same way making sure that the spaces between the teeth are clean.

Lay the balance spring on a piece of white paper. Take the cleaning brush with the soft bristles and charge it with chalk.

Hold the balance spring down with the tip of a finger and carefully dab the bristles of the brush into the coils of the exposed portion of the spring. Continue round until the complete spring has been cleaned.

The pivots, pivot bearings and pinions now need attention. Shave a point on the end of a piece of pegwood, insert it in the pivot bearings and lightly rotate it. This will ensure that all bearings are clear.

The point of the pegwood is now used to clean the pinions. Run the pegwood up and down each leaf removing any dirt that may be present.

The pivots are cleaned by pushing them into the end of a pith stick and rotating the pith a few times.

Having cleaned and polished all the parts we are now ready to assemble them.

Oil the ratchet wheels and clicks and fit them to the back plate. Place the balance and lever to one side and assemble the remaining parts to the back plate.

Lower the front plate into position gently guiding the pivots into their respective bearings. Make quite sure the wheel train is free-running, tighten down the front plate and again check the freedom of the wheel train.

The lever may now be fitted by slackening the front plate nuts, lifting the front plate slightly and inserting the lever into position. Lower the front plate again, check the lever for freedom and tighten the front plate nuts.

The mainspring may now be fully wound up and the clamp removed. Check the action of the wheel train and the escapement lever by operating the pallets a few times.

We are now ready to assemble the balance. Support the balance wheel by placing the lower pivot in a stake. Place the collet on the upper pivot and with a tubular punch gently tap the collet into position. By inserting the tool previously mentioned in the slot of the collet, the collet can be turned until it coincides with the mark made on the balance wheel during dismantling.

The position for the collet on the staff is determined by the position of the stud. The balance spring must lie quite flat when secured in the stud hole.

Fit the balance to the movement and adjust the upper bearing screw until end-shake of the balance staff disappears. Now unscrew the bearing screw slightly until there is just sufficient end-shake for free movement.

Rotate the balance until the impulse pin enters the lever and at the same time guide the end of the spring through the index pins and into the stud hole. The actual position of the spring to be pinned in the stud will be indicated by the kink in the spring caused by the previous pinning. Insert the pin into the stud and carefully squeeze it in tightly using the thin pliers. The balance should now be free to vibrate.

Having assembled the balance it is necessary to check the beat of the escapement.

Move the balance wheel round with the tip of a finger and watch the action of the escape wheel. Immediately a tooth of the escape wheel falls away from a pallet, stop the movement of the balance wheel and note its position. Now reverse the direction of the balance wheel and continue moving it until a tooth of the escape wheel falls away from the other pallet. Stop moving the balance wheel and note its new position. If the escapement beat is correct, the balance wheel will have moved the same distance on each side.

Rest the centre arbor on a stake and fit the cannon pinion by gently tapping it down with a hollow punch. The alarm wheel is now placed in position and the alarm wheel pin fitted.

Oil the wheel train pivot bearings, the balance bearing screw sinks and the pallets. Put one drop of oil on each pinion. Oil the rim of the alarm wheel and the edges of the mainspring and the alarm spring.

Now fit the hour wheel and the minute wheel and secure the dial. Fit the hand-set button and the alarm-set button and screw on the two winding keys.

We now have to synchronize the alarm mechanism with the time mechanism. Remove the tension from the alarm spring clamp by further winding the spring and then remove the clamp.

Slowly turn the alarm-set button until the alarm wheel pin drops into the base of the cam. This is the position at which the alarm will operate. The alarm hand is now fitted so that it exactly points to one particular hour, say 9 o'clock. The clock hands must therefore indicate the same time. Fit the hour hand to 9 and the minute hand to 3 minutes before the 12. To check the setting turn the alarm-set hand to ring at 10 o'clock and then operate the hand-set until the hands point to four minutes before 10 o'clock. Leave the clock to function and note the time at which the alarm operates. Adjustments can be made by removing the minute hand and repositioning it. The movement is now ready for fitting back into its case.

There is an alternative method of cleaning which is quicker but not so reliable. The movement is taken from its case and the hands, dial balance and the lever are removed.

The mainspring and the alarm spring are wound up and the movement is totally immersed in the cleaning fluid during the running down of the two springs. The movement is then lifted out of the fluid and a brush is used to clean the pivot holes in the plates. The springs are re-wound and the movement once again immersed in the cleaning fluid.

When the springs are fully run down, the movement is removed surplus fluid is shaken off, the plates are wiped with a clean piece of linen and the movement hung up to dry.

The balance and the lever are treated as previously described and are refitted to the movement when the movement is completely dry. Oiling is carried out as before.

CHAPTER TWENTY-FOUR

Electric clocks

AN electric clock consists of an electric motor, powered by an A.C. current, driving a pair of hands by means of reduction gearing. The speed of the electric motor will vary with the make and also with the electrical supply for which it was designed. An average speed is about 200 revolutions per minute.

In Chapter 4 we said that most wheel or gear trains are designed to reduce speed and increase power but that in mechanically powered time-pieces the reverse is the case.

Now that we are dealing with an electrically powered time-piece we find that the gear train is used in its conventional form, i.e., to provide a reduction of speed.

The ratio of the reduction must be such as to provide two gears with speeds of one revolution per hour and one revolution per twelve hours, respectively. These then will be the gears to carry the hands.

Assuming the movement has been removed from its case, take off the electric motor and examine the terminals and wires for security and cleanliness. If necessary resolder the wires to make good contact. Use a resin flux when soldering. Do not use spirits of salts because of its corrosive properties and the problems of washing it off.

Having made sure that the electrical connexions are good, try the gear train for freedom. If all the gears spin freely with no signs of harshness then refit the motor.

Support the movement over a cardboard box and connect up to the mains supply. If the motor fails to operate then it will have to be overhauled.

Some motors are sealed by the manufacturers with the intention of providing a replacement unit service. Unless you are skilled in the overhaul of electric motors it is better to return the unit to the manufacturer or one of his agents.

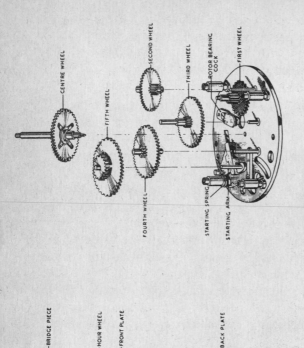

CENTRE WHEEL

FIFTH WHEEL

SECOND WHEEL

THIRD WHEEL

FOURTH WHEEL

ROTOR BEARING COCK

FIRST WHEEL

STARTING SPRING

STARTING ARM

ROTATION CONTROL
MECHANISM ON
SELF START MOVEMENT

BRIDGE PIECE

HOUR WHEEL

FRONT PLATE

BACK PLATE

STATOR COMPLETE

MINUTE PINION

MINUTE WHEEL

HAND SET

ROTOR

Fig. 91.

The rest of the movement can be dismantled and washed in petrol. Polish all the parts by the chalk-brushing method and clean out pivot holes with pegwood. Examine the gears for broken or damaged teeth particularly the fibre wheel. Assemble the movement and oil all bearing surfaces. The oil should be the grade supplied for electric clocks.

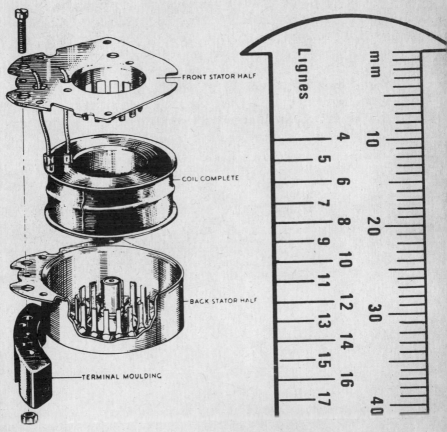

Fig. 92.

Fig. 93. Ligne gauge

The illustrations accompanying this chapter are those of the well-known Bijou movement manufactured by Smiths Clocks & Watches, Ltd. of England. The construction of the motor can be clearly seen as well as the layout of the train of gears.

The majority of electric clocks are fitted with motors that synchronize with the alternators at the power station. These alternators run at a controlled frequency which enables very accurate timekeeping to be maintained.

Occasionally the frequency of the alternators might drop a little at peak periods which will cause the clock to lose. In this event no adjustment should be made on the hand-set because the power station will automatically regulate the situation by increasing the speed of the generators when the peak period is over.

When the movement is finally assembled connect it to the electrical supply and check it before fitting it back into its case.

It must be remembered here that the hands will not commence moving straight away as a few seconds must elapse whilst the backlash of the gears is being taken up.

Mail ordering

The first thing to know is where to write to obtain tools and materials, and where to send repair work.

A study of the advertisements in the leading horological journals will provide this information but a few dealers' names and addresses have been included at the end of this appendix.

The official journal of the American Watchmakers Institute is the American Horologist & Jeweler. Other leading horological journals are listed.

Select a few suppliers and repairers and write asking for copies of their illustrated catalogs of tools and materials and for details of their postal repair service. Catalogs are well worth having.

In the world of horology spare parts for watch and clock movements are known as materials, and those who supply them are material dealers.

Many repairers supply order forms printed in tabulated form listing those parts of the movement that are required to be forwarded when ordering specific materials. Similar instructions are given for repair work.

Take care to carry out the instructions on the order form. This will prevent mistakes and save time.

Great care must be exercised in preparation for postage. Individual parts and assemblies should be wrapped separately in tissue paper. These should be placed in a metal or strong cardboard box.

Crumpled tissue paper can then be used to fill the space and prevent movement.

Wrap the order form around the box and, if the first container is very small, place it in a larger container filling the space with crumpled newspaper.

When ordering material give as much information as you can. Always state the caliber number of a watch movement which will be stamped on one of the plates. Additional information could be manufacturers' name, brand name on dial, diameter of watch front plate in lignes or millimeters, center sweep, small seconds, self-winding (Fig 93), water resistant, type of clock such as Westminster chime, striking, grandfather, weight or spring powered, cuckoo, travel-alarm, carriage etc. It may even be necessary to send the movement.

When working at the bench it is easy to drop a part or it may even spring from the tweezers. Because we are working with such small pieces they are not easily found. A carpeted floor makes it almost impossible. With a smooth polished floor you stand a better chance of finding it.

Clicksprings, set springs and small screws are the biggest offenders. When fitting a small spring it is advisable to cover it with one of your fingers in case it jumps and is lost. If you do lose one then when writing for a replacement ask for a gross of assorted sizes. They are not expensive and who knows, you may need another before the job is finished.

Finally wrap the package in strong wrapping paper, seal with gummed sealing tape, and send by registered mail.

Always retain the containers forwarded by the supplier. These containers are most useful and can be used repeatedly.

Horological Journals

American Horoligst & Jeweler   2403 Champa Street
                     (monthly)   Denver
                                   Colorado 80205

British Jeweller   (monthly) Available from:—
Canadian Jeweller   (”   ) A. A. Osborne & Son
Horological Journal   (”   ) c/o House of Clocks
Watchmaker, Jeweller &   707 So. Hill Street
  Silversmith   (”   ) Los Angeles
Watchmakers of Victoria   California 90014
  (Australian)   (”   )
Jeweller & Metalworker   (bi-monthly)

List of Suppliers

The following addresses of suppliers and repairers have been taken from ads in the American Horologist & Jeweler. The key is as accurate as the advertisements will permit, but no doubt many of the suppliers carry a wider range than is indicated, e.g., tool suppliers may also stock watch and clock parts.

A Clock movements
B   "   materials
C   "   repairs
D   "   dial repairs and refinishing
E   "   parts made (Wheel cutting)
F   "   springs
G Watch case repairs
H   "   movements (used)
I   "   repairs
J   "   band repairs

K Watch materials
L Tools general
M   "   precision and lathes
N Cleaning and timing machine repairs
O Catalogs or lists
P General supplier
Q Workbenches
R Short lengths of metal
S Part exchange equipment
T Synchronized motors and timer parts

Antique Nook, The, Inc. D,F,O
Box 338
Atwater, Ohio 44201

Becker-Heckman Company G
5 N. Wabash Ave. Rm. 1012
Chicago, Ill. 60602

Benders Watch Service I
205 E. Ashdale St.
Philadelphia, Pa. 19120

Borel, Jules & Company K
1110 Grand
Kansas City, Mo. 64106

Borel & Frei, Inc. K
315 W. Fifth St.
Los Angeles, Calif. 90013

Campbell M,R
1424 Barclay
Springfield, Ohio 45505

Cleveland-Miami Band Repair Co. J
P.O. Box 1259
St. Petersburg, Fla. 33731

Davidson Jewelers Supplies L
861 - 6th Ave. Rm. 310
San Diego, Calif. 92101

Donahue, Stanley Co., Inc. L
705 Main Street Rm. 530
Houston, Texas 77002

Empire Clock Company B,O,S
1295 Rice Street
St. Paul, Minn. 55117

Frei, Otto-Borel, Jules, Inc. K
P.O. Box 796
Oakland, Calif. 94606

R. Givler C
146 3rd
Wadsworth, Ohio 44281

Goldfarb, A. J., Inc.   O
251 W. 30th St.
New York, N.Y. 10001

Green's Jewelers Supply   L
205 Burk Burnett Bldg.
4th & Main Streets
Forth Worth, Texas 76102

Hamlin Clock Dials   D
299 S. Vinedo Avenue
Pasadena, Calif. 91107

Henderson, Don Company   N
Box 1260
Ada, Okla. 74820

Herman, Ralph Clock
   House   B,C,O
628 Coney Island Ave.
Brooklyn, N.Y. 11218

Hersey Watch Timer
   Service   N
226 S. Wabash Ave.
Chicago, Ill. 60604

Horolovar Company, The   B,O
Box 400 A
Bronxville, N.Y. 10708

Iowa Jewelry Supply Co.   L
317-320 Empire Building
6th & Walnut Streets
Des Moines, Ia. 50309

K & J Watch Service   I
P.O. Box 9262
Denver, Colorado 80209

Kilb & Company   O
623 N. 2nd Street
P.O. Drawer 8-A
Milwaukee, Wisc. 53201

LaRose, S., Inc.   A,B,L,O
Greensboro, No. Car 27420

Marshall, C & E   L
1113 West Belmont
Chicago, Ill. 60657

Mason & Sullivan Co.   A,D,O
Osterville, Mass. 02655

Moses, Bob (The
   Watchmaker)   C,I,O
Box 1
Pleasantville, N.Y. 10570

Nordman & Aurich   L
657 Mission Street
San Francisco, Calif. 94105

Osborne, A. A. & Son   B,E
c/o House of Clocks
707 So. Hill Street
Los Angeles, Calif. 90014

Rosenthal Jewelers Supply
   Corp.   L
139 N.E. 1st Street - Suite 620
Miami, Fla. 33132

St. Louis Refining
   Company   C,H,I,O,P
802 Arcade Building
812 Olive St.
St. Louis, Mo. 63101

Smelting & Refining Co.,
   Inc.   Q
P.O. Box 2010
1712 Jackson
Dallas, Texas 75221

Swigart, E. & J. Co.   B,K
34 W. 6th Street
Cincinnati, Ohio 45202

Tela-Time Industries Tronix
   Toll Corp.   L,S
37 West 47th Street
New York, N.Y. 10036

Tick Tock Shop   C,I
P.O. Box 44181
Cleveland, Ohio 44144

Wade, H. R. & Son, Inc.   L
608 East Central
Wichita, Kansas 67202

Wisconsin Jewelers
   Supply Co.   L
Madison Bldg.
623 N. 2nd Street
Milwaukee, Wisc. 53203

Index

INDEX